Troubleshooting

Analog Circuits

Troubleshooting
Analog Circuits

With Electronics Workbench Circuits

Robert A. Pease
and
Interactive Image Technologies

Newnes
An Imprint of Elsevier

Boston   Oxford   Johannesburg   Melbourne   New Delhi   Singapore

Newnes
An Imprint of Elsevier
Copyright © 1991 by Elsevier
Paperback reprint 1993

 This book is printed on acid-free paper.

ISBN-13: 978-0-7506-9499-5
ISBN-10: 0-7506-9499-8

The publisher offers special discounts on bulk orders of this book.
For information, please contact:
Manager of Special Sales
Elsevier
200 Wheeler Road
Burlington, MA 01803
Tel: 781-313-4700
Fax: 781-313-4802

For information on all Newnes publications available, contact our World Wide
Web homepage at http://www.newnespress.com

Transferred to Digital Printing, 2011

Printed and bound in the United Kingdom

Contents

Contents

Foreword

"Your idea is so good that, if you give me 20 minutes, I'll be sure that I was the first one to think of it." Although I pass out that accolade sparingly, if I were to do what the compliment implies, I'd surely claim credit for the idea of publishing Bob Pease's series on "Troubleshooting Analog Circuits" in *EDN* Magazine Edition. The fact is, though, that the idea came from Jon Titus, VP, Editorial Director, and Chief Editor of *EDN* magazine and from Tarlton Fleming, then an *EDN* Associate Editor and now Manager of Applications Engineering at Maxim Integrated Products Corporation.

In early 1988, Jon and those *EDN* technical editors who work at the publication's (and Cahners Publishing Company's) Newton, Massachusetts, headquarters were brainstorming ideas for articles we could solicit from contributors in industry. Jon ventured that because *EDN* readers always look to the magazine to provide practical ideas on how to do their jobs better, and because trouble is ubiquitous, articles on how to troubleshoot more effectively should be a natural for us.

Tarlton, who edited *EDN*'s popular Design Ideas section, worked with Bob on a regular basis, as Bob reviews the analog design ideas submitted by *EDN* readers. Tarlton recalled Bob's mentioning a book he and his colleagues at National Semiconductor were planning to write on power-supply design. Tarlton said he thought Bob had already put together some material on troubleshooting. We needed to find out whether National would grant *EDN* the rights to publish a portion of the book. Tarlton would open the discussions.

Shortly afterward, a good-sized package arrived at *EDN*'s offices. In it was the text of what would eventually become the first three installments of Bob's series. By then, Tarlton had left the East Coast to seek fame and fortune in Silicon Valley, so the task of reviewing Bob's material fell to me. I skimmed through it quickly and became quite intrigued.

I am a contemporary of Bob's; actually, I am a few years older. Though we did not know each other at the time, I was a graduate student in EE at MIT while Bob was an undergraduate there. I first became aware of Bob when he was working for his previous employer, George A. Philbrick Researches, now a part of Teledyne Components in Dedham, Massachusetts. Even in those days—the sixties and early seventies—Bob was a prolific writer. He shared his musings and technical insights with Philbrick customers and other analog engineers who read the firm's house organ, "The Lightning Empiricist," and with readers of trade magazines, such as *EDN*.

Those earlier writings did a lot to burnish Bob's image as a technical expert, but they had a secondary effect as well: They made his sense of humor and his passion for puns something of a legend. As a form of humor, plays on words are denigrated by all too many people. However, at least a few openly admit to enjoying puns, and that group includes Bob and myself. Many years ago, when I first read material Bob had written, I suspected that if I ever met him, I'd probably like him. When I started to read what he had just submitted to *EDN*, the experience was a bit like a chance encounter with an old friend after not meeting up with him for a long time.

The material was somewhat out of the ordinary for *EDN*. It was technical, yes ... but it was lighter than most of what we publish. There were few equations and no

Acknowledgments

I would like to dedicate this book to my old friend Bruce Seddon. Starting 30 years ago, he helped me appreciate some of the niceties of worst-case design. They never did teach that at school, so you have to have a wise old-timer to learn it from. Bruce was never too busy to lend an ear and a helping hand, and if I never got around to saying thank-you—well, 30 years is a long time to be an ingrateful lazy bum, but now's the time to say, "Thank you, Bruce."

I want to express my appreciation to the 40-odd friends who helped review the drafts of these articles, correct my mistakes, and suggest additions. Special thanks go to Jim Moyer, Tim Regan, Dennis Monticelli, Larry Johnson and to Dan Strassberg at *EDN*, who contributed significant technical ideas that were beyond my experience. I also want to thank Cindy Lewis of Sun Circuits Inc. (Santa Clara, CA.) for her help in preparing the table of PC-board materials in Chapter 5. Credit goes to Mineo Yamatake for his elegant thermocouple amplifier design, Steve Allen, Peggi Willis, Al Neves, and Fran Hoffart for their photography, and Erroll Dietz as Key Grip and Carlos Huerta as Gaffer. Thanks also to Hendrick Santo and to the people at Natasha's Attic in San Jose, for their help in engineering, assembling, and styling the Czar's Uniform. And kudos to each of the *EDN* editors who slaved over my copy: Julie Anne Schofield, Anne Watson Swager, Charles H. Small, and Dan Strassberg, in addition to Carol S. Lewis at HighText Publications in San Diego. Each one really worked hard, and cared a lot about every word and phrase that we debated and argued and polished and refined.

I am also grateful to Joyce Gilbert, our group's secretary, who wound up typing a lot more than she bargained for. She believed me when I told her it would only be 50 or 60 pages typed . . . how were we to know how this would grow to 280 pages?? Note, even though Joyce typed everything in here, I typed everything, too, as I find that my creative juices flow best when I am typing on a good word processor. I wouldn't have asked her to type anything I wouldn't type. However, with the price of computers being what they are, shrinking and lowering, I would never want to require anybody to re-key this kind of text, never again. It's not that expensive or difficult to type in an ASCII-compatible format in the first place. I typed my early draft on my old Coleco ADAM, with a non-compatible cassette memory. Joyce re-typed everything I wrote with Ashton-Tate Multi-mate, and we sent the ASCII files to *EDN*, back in August and November of 1988. I got the typed files back from *EDN* and put in dozens and dozens of hours retyping, polishing, refining, and expanding the text. I also want to express my appreciation for Wanda Garrett, who put up with an awful lot of dumb questions about how to get the word-processor running for me. If any of my readers is ever going to write a book, well, think about what you are going to do, and how you are going to do it. Remember, this text started out as a single chapter for Al Kelsch's book on switching regulators! I wouldn't have gone about this in such a dumb, inefficient way if I could have imagined what a big project it would be. But, then, I might never have even started . . . .

As for technical and troubleshooting ideas, well, after all the tips I've given you, it's only fair that you share your comments with me!

Bob Pease, Staff Scientist
National Semiconductor Corp., M/S C2500A, P.O. Box 58090, Santa Clara, CA 95052-8090.

About the Author

For the record, Bob Pease is a senior scientist in industrial linear-IC design at National Semiconductor Corporation in Santa Clara, California; he has worked at National since 1976. He is also one of the best-known analog-circuit designers in the world—he's been creating practical, producible analog products for fun (his) and profit (both his and his employers') and writing about analog topics for over a quarter of a century.

As you might expect, though, there's a lot more to Bob Pease than his impressive credentials. Following untrodden paths to discover where they lead is one of Bob's avocations. He's done it on foot, on skis, and on a bicycle—sometimes by himself and sometimes with his wife and two sons—often along abandoned railroad roadbeds throughout the United States and England. Aside from the peace and quiet and the thrill of the journey itself, the reward for these wanderings is observing vistas of America that few people have seen. The curiosity that motivates Bob's exploration of old railroad routes is reflected in many of his other activities both at and away from work.

For example, another of Bob's hobbies is designing voltage-to-frequency converters (VFCs). Most people who design VFCs do it as part of a job. Although Bob sometimes designs VFCs for use in National products, he often does it just for fun and because he finds the activity educational and challenging. A couple years ago, on such a lark, he put together a VFC that used only vacuum tubes. This circuit proved that the company where he spent the first 14 years of his career, George A. Philbrick Researches, (more recently, Teledyne-Philbrick, now Teledyne Components of Dedham, MA) could have gone into the VFC business in 1953—eight years before Pease received his BSEE from MIT. Twenty years after he designed it, one of Bob's first solid-state VFCs, the 4701, continues to sell well for Teledyne-Philbrick. The story of how Pease pioneered the voltage-to-frequency business is recounted in a chapter of *Analog Circuit Design: Art, Science, and Personalities* (Butterworth-Heinemann, 1991), edited by Jim Williams. (See the ad at the end of this book.)

Bob also loves to write—he clearly enjoys communicating to others the wisdom he has acquired through his work. He has published about 60 magazine articles (not counting the series in *EDN* that led to this book) and holds approximately ten US patents. Recently he began a series of columns in *Electronic Design* magazine, where he comments fortnightly on various aspects of linear and analog circuits.

Bob takes great delight in seeing his ideas embodied in the work of others. For example, one of his proudest accomplishments is a seismic preamplifier that he designed for an aerospace company during his coffee break. After many years of service, the amplifier was still at work on the moon, amplifying and telemetering moon-quakes (but its batteries may have recently expired). Bob also designed a compact 1/3-ounce voltage-to-frequency module that was carried to the summit of Mt. Everest, where it was used to convert medical and scientific data for medical research, with the 1980 American Medical Research Expedition (from the University of California Medical School at La Jolla).

National has taken advantage of Bob's penchant for providing ideas that others can use. In his role of senior scientist, Bob's responsibilities—besides designing voltage references and regulators, temperature sensors, and VFC ICs—include consulting with co-workers, fielding applications questions that have stumped other engineers, and reviewing colleagues' designs. In a similar vein, Bob is a long-time contributing editor who reviews design-idea submissions of analog circuits for *EDN* magazine.

1. First Things First

The Philosophy of Troubleshooting

In this first chapter, I will make the point that a significant part of effective trouble-shooting lies in the way that you think about the problem. The next chapter will cover the equipment you should buy and build to help you diagnose problems. Other chapters will illuminate some of the more subtle and elusive characteristics of passive and active components, and the PC boards and cables that interconnect them.

Troubleshooting Is More Effective with the Right Philosophy

If you recall that the most boring class in school was a philosophy class, and you think this book will be boring that way, well, WRONG. We are going to talk about the real world and examples of mistakes, goofs, and how we can recover from these mistakes. We are going to talk about all the nasty problems the world tries to inflict on us. We are talking about Trouble with a capital T, and how to overcome it.

Here at National Semiconductor, we decided a couple of years ago that we ought to write a book about switch-mode power supplies. Within the applications and design groups, nearly all of the engineers volunteered to write a chapter, and I volunteered to do a chapter on troubleshooting. At present, the status of that book is pretty dubious. But, the "troubleshooting chapter" is going strong, and you readers are among the first to benefit, because that one chapter has expanded to become this entire book. Although I am probably not the world's best analog-circuit trouble-shooter, I am fairly good; and I just happened to be the guy who sat down and put all these stories in writing.

Furthermore, the techniques you need to troubleshoot a switch-mode power supply apply, in general, to a lot of other analog circuits and may even be useful for some basic digital hardware. You don't have to build switchers to find this book useful; if you design or build any analog circuits, this book is for you.

Maybe there are some engineers who are knowledgeable about digital circuits, computers, microprocessors, and software, who may someday write about the trou-bleshooting of those types of circuits. That sure would suit me fine, because I am certainly not going to talk about those circuits!! Everybody has to be ignorant about something, and *that* is exactly what I am ignorant about.

If Only Everything Would Always Go Right . . .

Why are we interested in troubleshooting? Because even the best engineers take on projects whose requirements are so difficult and challenging that the circuits don't work as expected—at least not the first time. I don't have data on switching regula-tors, but I read in an industry study that when disk drives are manufactured, the frac-tion that fails to function when power is first applied typically ranges from 20 to 70%. Of course, this fraction may occasionally fall as low as 1% and rise as high as 100%. But, on the average, production engineers and technicians must be prepared to repair 20, 40, or 60% of these complex units. Switching-regulated power supplies can also be quite complex. If you manufacture them in batches of 100, you shouldn't be surprised to find some batches with 12 pieces that require troubleshooting and other batches that have 46 such pieces. The troubleshooting may, as you well know, be tough with a new product whose bugs haven't been worked out. But it can be even tougher when the design is old and the parts it now uses aren't quite like the ones you used to be able to buy. Troubleshooting can be tougher still when there isn't much documentation describing how the product is supposed to work, and the designer isn't around any more. If there's ever a time when troubleshooting *isn't* needed, it's just a temporary miracle. You might try to duck your troubleshooting for a while. You might pretend that you can avoid the issue.

And, what if you decide that troubleshooting isn't necessary? You may find that your first batch of products has only three or four failures, so you decide that you don't need to worry. The second batch has a 12% failure rate, and most of the rejects have the same symptoms as those of the first batch. The next three batches have failure rates of 23, 49, and 76%, respectively. When you finally find the time to study the problems, you will find that they would have been relatively easy to fix if only you had started a couple of months earlier. That's what Murphy's Law can do to you if you try to slough off your troubleshooting chores...we have all seen it happen.

If you have a bunch of analog circuits that you have to troubleshoot, well, why don't you just look up the troubleshooting procedures in a book? The question is excellent, and the answer is very simple: Until now, almost nothing has been written about the troubleshooting of these circuits. The best previous write-up that I have found is a couple pages in a book by Jiri Dostal (Ref. 1). He gives some basic procedures for looking for trouble in a fairly straightforward little circuit: a voltage reference/regulator. As far as Dostal goes, he does quite well. But, he only offers a few pages of troubleshooting advice, and there is much to explain beyond what he has written. [1]

Another book that has several good pages about the philosophy of troubleshooting is by John I. Smith (Ref. 2). Smith explains some of the foibles of wishing you had designed a circuit correctly when you find that it doesn't work "right." Unfortunately, it's out of print. Analog Devices sells a Data Converter Handbook (Ref. 3), and it has a few pages of good ideas and suggestions on what to look for when troubleshooting data converter and analog circuits.

What's missing, though, is general information. When I started writing about this troubleshooting stuff, I realized there was a huge vacuum in this area. So I have filled it up, and here we are.

You'll probably use general-purpose test equipment. What equipment can you buy for troubleshooting? I'll cover that subject in considerable detail in the next chapter. For now, let me observe that if you have several million dollars worth of circuits to troubleshoot, you should consider buying a $100,000 tester. Of course, for that price you only get a machine at the low end of the line. And, after you buy the machine, you have to invest a lot of time in fixturing and software before it can help you. Yes, you can buy a $90 tester that helps locate short circuits on a PC board; but, in the price range between $90 and $100,000, there isn't a lot of specialized troubleshooting equipment available. If you want an oscilloscope, you have to buy a general-purpose oscilloscope; if you want a DVM, it will be a general-purpose DVM.

1. I must say, I recently re-read Mr. Dostal's book, and it is still just about the best technical book on operational amplifiers. It's more complete, more technical, but less intuitive than Tom Frederiksen's *Intuitive IC Op Amps*. Of course, for $113, it *ought* to be pretty good. It is getting a little old and dated, and I hope he plans to update it with a new revision soon.

Now, it's true that some scopes and some DVMs are more suitable for trouble-shooting than others (and I will discuss the differences in the next chapter), but, to a large extent, you have to depend on your wits.

Your wits: Ah, very handy to use, your wits—but, then what? One of my favorite quotes from Jiri Dostal's book says that troubleshooting should resemble fencing more closely than it resembles wrestling. When your troubleshooting efforts seem like wrestling in the mud with an implacable opponent (or component), then you are probably not using the right approach. Do you have the right tools, and are you using them correctly? I'll discuss that in the next chapter. Do you know how a failed component will affect your circuit, and do you know what the most likely failure modes are? I'll deal with components in subsequent chapters. Ah, but do you know how to think about Trouble? That is this chapter's main lesson.

Even things that can't go wrong, do. One of the first things you might do is make a list of all the things that could be causing the problem. This idea can be good up to a point. I am an aficionado of stories about steam engines, and here is a story from the book *Master Builders of Steam* (Ref. 4). A class of new 3-cylinder 4-6-0 (four small pilot wheels in front of the drive wheels, six drive wheels, no little trailing wheels) steam engines had just been designed by British designer W. A. Stanier, and they were "... perfect stinkers. They simply would not steam." So the engines' designers made a list of all the things that could go wrong and a list of all the things that could not be at fault; they set the second list aside.

The designers specified changes to be made to each new engine in hopes of solving the problem: "Teething troubles bring modifications, and each engine can carry a different set of modifications." The manufacturing managers "shuddered as these modified drawings seemed to pour in from Derby (Ed: site of the design facility—the Drawing Office), continually upsetting progress in the works." (Lots of fun for the manufacturing guys, eh?) In the end, the problem took a long time to find because it was on the list of "things that couldn't go wrong."

Allow me to quote the deliciously horrifying words from the text: "Teething troubles always present these two difficulties: that many of the clues are very subjective and that the 'confidence trick' applies. By the latter I mean when a certain factor is exonerated as trouble-free based on a sound premise, and everyone therefore looks elsewhere for the trouble: whereas in fact, the premise is not sound and the exonerated factor is guilty. In Stanier's case this factor was low super-heat. So convinced was he that a low degree of super-heat was adequate that the important change to increased superheater area was delayed far longer than necessary. There were some very sound men in the Experimental Section of the Derby Loco Drawing Office at that time, but they were young and their voice was only dimly heard. Some of their quite painstaking superheater test results were disbelieved." But, of course, nothing like that ever happened to anybody you know—right?

Experts Have No Monopoly on Good Advice

Another thing you can do is ask advice only of "experts." After all, only an expert knows how to solve a difficult problem—right? Wrong! Sometimes, a major reason you can't find your problem is because you are too close to it—you are blinded by your familiarity. You may get excellent results by simply consulting one or two of your colleagues who are not as familiar with your design; they may make a good guess at a solution to your problem. Often a technician can make a wise (or lucky) guess as easily as can a savvy engineer. When that happens, be sure to remember who saved your neck. Some people are not just "lucky"—they may have a real knack for solving tricky problems, for finding clues, and for deducing what is causing the trouble. Friends like these can be more valuable than gold.

At National Semiconductor, we usually submit a newly designed circuit layout to a review by our peers. I invite everybody to try to win a Beverage of Their Choice by catching a real mistake in my circuit. What we *really* call this, is a "Beercheck." It's fun because if I give away a few pitchers of brew, I get some of my dumb mistakes corrected—mistakes that I myself might not have found until a much-later, more-painful, and more-expensive stage. Furthermore, we all get some education. And, you can never predict who will find the little picky errors or the occasional real killer mistake. All technicians *and* engineers are invited.

Learn to Recognize Clues

There are four basic questions that you or I should ask when we are brought in to do troubleshooting on someone else's project:

- Did it *ever* work right?
- What are the symptoms that tell you it's not working right?
- When did it start working badly or stop working?
- What other symptoms showed up just before, just after, or at the same time as the failure?

As you can plainly see, the clues you get from the answers to these questions might easily solve the problem right away; if not, they may eventually get you out of the woods. So even if a failure occurs on your own project, you should ask these four

Figure 1.1. Peer review is often effective for wringing problems out of designs. Here, the author gets his comeuppance from colleagues who have spotted a problem *because* they are not as overly familiar with his circuit layout as he is. (Photo by Steve Allen.)

questions—as explicitly as possible—of yourself or your technician or whoever was working on the project. For example, if your roommate called you to ask for a lift because the car had just quit in the middle of a freeway, you would ask whether anything else happened or if the car just died. If you're told that the headlights seemed to be getting dimmer and dimmer, that's a *clue*.

Ask Questions; Take Notes; Record Amount of Funny

When you ask these four questions, make sure to record the answers on paper—preferably in a notebook. As an old test manager I used to work with, Tom Milligan, used to tell his technicians, "When you are taking data, if you see something funny, Record Amount of Funny." That was such a significant piece of advice, we called it "Milligan's Law." A few significant notes can save you hours of work. Clues are where you find them; they should be saved and savored.

Ask not only these questions but also any other questions suggested by the answers. For example, a neophyte product engineer will sometimes come to see me with a batch of ICs that have a terrible yield at some particular test. I'll ask if the parts failed any other tests, and I'll hear that nobody knows because the tester doesn't continue to test a part after it detects a failure. A more experienced engineer would have already retested the devices in the RUN ALL TESTS mode, and that is exactly what I instruct the neophyte to do.

Likewise, if *you* are asking another person for advice, you should have all the facts laid out straight, at least in your head, so that you can be clear and not add to the confusion. I've worked with a few people who tell me one thing and a minute later start telling me the opposite. Nothing makes me lose my temper faster! Nobody can help you troubleshoot effectively if you aren't sure whether the circuit is running from +12 V or ±12 V and you start making contradictory statements.

And, if I ask when the device started working badly, don't tell me, "At 3:25 PM." I'm looking for clues, such as, "About two minutes after I put it in the 125 °C oven," or, "Just after I connected the 4 Ω load." So just as we can all learn a little more about troubleshooting, we can all learn to watch for the clues that are invaluable for fault diagnosis.

Methodical, Logical Plans Ease Troubleshooting

Even a simple problem with a resistive divider offers an opportunity to concoct an intelligent troubleshooting plan. Suppose you had a series string of 128 1 kΩ resistors. (See Figure 1.2.) If you applied 5 V to the top of the string and 0 V to the bottom, you would expect the midpoint of the string to be at 2.5 V. If it weren't 2.5 V but actually 0 V, you could start your troubleshooting by checking the voltage on each resistor, working down from the top, one by one. But that strategy would be absurd! Check the voltage at, say, resistor #96, the resistor which is halfway up from the midpoint to the top. Then, depending on whether that test is high, low, or reasonable, try at #112 or #80—at 5/8 or 7/8 of the span—then at #120 or #104 or #88 or #72, branching along in a sort of binary search—that would be much more effective. With just a few trials (about seven) you could find where a resistor was broken open or shorted to ground. Such branching along would take a lot fewer than the 64 tests you would need to walk all the way down the string.

Further, if an op-amp circuit's output were pegged, you would normally check the circuit's op amp, resistors, or conductors. You wouldn't normally check the capacitors, *unless* you guessed that a shorted capacitor could cause the output to peg.

Conversely, if the op amp's V_{OUT} was a few dozen millivolts in error, you might start checking the resistors for their tolerances. You might not check for an open-circuited or wrong-value capacitor, *unless* you checked the circuit's output with a scope and discovered it oscillating!! So, in any circuit, you must study the data—your "clues"—until they lead you to the final test that reveals the true cause of your problem.

Thus, you should always first formulate a hypothesis and then invent a reasonable test or series of tests, the answers to which will help narrow down the possibilities of what is bad, and may in fact support your hypothesis. These tests should be performable. But you may define a test and then discover it is not performable or would be much too difficult to perform. Then I often think, "Well, if I could do that test, the answer would either come up 'good' or 'bad.' OK, so I can't easily run the test. But if I assume that I'd get one or the other of the answers, what would I do next to nail down the solution? Can I skip to the next test??"

For example, if I had to probe the first layer of metal on an IC with two layers of metal (because I had neglected to bring an important node up from the first metal to the second metal), I might do several other tests instead. I would do the other tests hoping that maybe I wouldn't have to do that probing, which is rather awkward even if I can "borrow" a laser to cut through all the layers of oxide. If I'm lucky, I may never have to go back and do that "very difficult or nearly impossible" test.

Of course, sometimes the actual result of a test is some completely unbelievable answer, nothing like the answers I expected. Then I have to reconsider—where were my assumptions wrong? Where was my thinking erroneous? Or, did I take my measurements correctly? Is my technician's data really valid? That's why troubleshooting is such a challenging business—almost never boring.

On the other hand, it would be foolish for you to plan everything and test nothing. Because if you did that, you would surely plan some procedures that a quick test would show are unnecessary. That's what they call "paralysis by analysis." All things being equal, I would expect the planning and testing to require equal time. If the tests are very complicated and expensive, then the planning should be appropriately comprehensive. If the tests are simple, as in the case of the 128 resistors in series, you could make them up as you go along. For example, the list above of resistors #80, 112, 120, 104, 88, or 72 are nominally binary choices. You don't have to go to exactly those places—an approximate binary search would be just fine.

You Can Make Murphy's Law Work *for* You

Murphy's Law is quite likely to attack even our best designs: "If anything can go wrong, it will." But, I can make Murphy's Law work *for* me. For example, according to this interpretation of Murphy's Law, if I drive around with a fire extinguisher, if I am prepared to put out any fire—will that make sure that I never have a fire in my car? When you first hear it, the idea sounds dumb. But, if I'm the kind of meticulous person who carries a fire extinguisher, I may also be neat and refuse to do the dumb things that permit fires to start.

Similarly, when designing a circuit I leave extra safety margins in areas where I cannot surely predict how the circuit will perform. When I design a breadboard, I often tell the technician, "Leave 20% extra space for *this* section because I'm not sure that it will work without modifications. And, please leave extra space around *this* resistor and *this* capacitor because I might have to change those values." When I design an IC, I leave little pads of metal at strategic points on the chip's surface, so that I can probe the critical nodes as easily as possible. To facilitate probing when

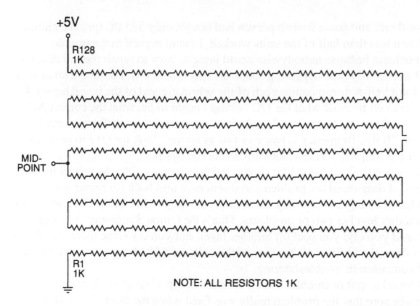

Figure 1.2. If you discovered that the midpoint was not at 2.5 V, but at 0 V, how would *you* troubleshoot this circuit? How would you search to detect a short, or an open?

working with 2-layer metal, I bring nodes up from the first metal through *vias* to the second metal. Sometimes I leave holes in my Vapox passivation to facilitate probing dice. The subject of testability has often been addressed for large digital circuits, but the underlying ideas of Design For Testability are important regardless of the type of circuit you are designing. You can avoid a lot of trouble by thinking about what can go wrong and how to keep it from going wrong before the ensuing problems lunge at you. By planning for every possibility, you can profit from your awareness of Murphy's Law. Now, clearly, you won't think of *every* possibility. (Remember, it was something that *couldn't* go wrong that caused the problems with Stanier's loco-motives.) But, a little forethought can certainly minimize the number of problems you have to deal with.

Consider Appointing a Czar for a Problem Area

A few years ago we had so many nagging little troubles with band-gap reference circuits at National, that I decided (unilaterally) to declare myself "Czar of Band Gaps." The main rules were that all successful band-gap circuits should be registered with the Czar so that we could keep a log book of successful circuits; all unsuccessful circuits, their reasons for failure, and the fixes for the failures should likewise be logged in with the Czar so that we could avoid repeating old mistakes; and all new circuits should be submitted to the Czar to allow him to spot any old errors. So far, we think we've found and fixed over 50% of the possible errors, before the wafers were fabricated, and we're gaining. In addition, we have added Czars for start-up circuits and for trim circuits, and a Czarina for data-sheet changes, and we are con-sidering other czardoms. It's a bit of a game, but it's also a serious business to use a game to try to prevent expensive errors.

I haven't always been a good troubleshooter, but my "baptism of fire" occurred quite a few years ago. I had designed a group of modular data converters. We had to ship 525 of them, and some foolish person had bought only 535 PC (printed circuit) boards. When less than half of the units worked, I found myself in the trouble-shooting business because nobody else could imagine how to repair them. I discovered that I needed my best-triggering scope and my best DVM. I burned a lot of midnight oil. I got half-a-dozen copies each of the schematic and of the board layout. I scribbled notes on them of what the DC voltages ought to be, what the correct AC waveforms looked like, and where I could best probe the key waveforms. I made little lists of, "If this frequency is twice as fast as normal, look for Q17 to be damaged, but if the frequency is low, look for a short on bus B."

I learned where to look for solder shorts, hairline opens, cold-soldered joints, and intermittents. I diagnosed the problems and sent each unit back for repair with a neat label of what to change. When they came back, did they work? Some did—and some still had another level or two of problems. That's the Onion Syndrome: You peel off one layer, and you cry; you peel off another layer, and you cry some more. . . . By the time I was done, I had fixed all but four of the units, and I had gotten myself one hell of a good education in troubleshooting.

After I found a spot of trouble, what did I do about it? First of all, I made some notes to make sure that the problem really was fixed when the offending part was changed. Then I sent the units to a good, neat technician who did precise repair work—much better than a slob like me would do. Lastly, I sent memos to the manu-facturing and QC departments to make sure that the types of parts that had proven troublesome were not used again, and I confirmed the changes with ECOs (Engineering Change Orders). It is important to get the paperwork scrupulously cor-rect, or the alligators will surely circle back to vex you again.

Sloppy Documentation Can End in Chapter XI

I once heard of a similar situation where an insidious problem was causing nasty reliability problems with a batch of modules. The technician had struggled to find the solution for several days. Finally, when the technician went out for lunch, the design engineer went to work on the problem. When the technician came back from lunch, the engineer told him, "I found the problem; it's a mismatch between Q17 and R18. Write up the ECO, and when I get back from lunch I'll sign it." Unfortunately, the good rapport between the engineer and the technician broke down: there was some miscommunication. The technician got confused and wrote up the ECO with an incorrect version of what should be changed. When the engineer came back from lunch, he initialed the ECO without really reading it and left for a two-week vacation.

When he came back, the modules had all been "fixed," potted, and shipped, and were starting to fail out in the field. A check of the ECO revealed the mistake—too late. The company went bankrupt. It's a true story and a painful one. Don't get sloppy with your paperwork; don't let it happen to you.

Failure Analysis?

One of the reasons you do troubleshooting is because you may be required to do a Failure Analysis on the failure. That's just another kind of paperwork. Writing a report is not always fun, but sometimes it helps clarify and crystallize your under-standing of the problem. Maybe if a customer had forced my engineer friend to write exactly what happened and what he proposed to do about it, that disaster would not have occurred. When I have nailed down my little problem, I usually write down a scribbled quick report. One copy often goes to my boss, because he is curious why it's been taking me so long. I usually give a copy to friends who are working on similar projects. Sometimes I hang a copy on the wall, to warn *all* my friends. Sometimes I send a copy to the manufacturer of a component that was involved. If you communicate properly, you can work to avoid similar problems in the future.

Then there are other things *you* can do in the course of *your* investigation. When you find a bad component, don't just throw it in a wastebasket. Sometimes people call me and say, "Your ICs have been giving me this failure problem for quite a while." I ask, "Can you send me some of the allegedly bad parts?" And they reply, "Naw, we always throw them in the wastebasket . . ." Please don't do that, because often the ability to troubleshoot a component depends on having several of them to study. Sometimes it's even a case of "NTF"—"No Trouble Found." That happens more often than not. So if you tell me, "Pease, your lousy op amps are failing in my circuit," and there's actually nothing wrong with the op amps, but it's really a misapplication problem—I can't help you very well if the parts all went in the trash. Please save them, at least for a while. Label them, too.

Another thing you can do with these bad parts is to open them up and see what you can see inside. Sometimes on a metal-can IC, after a few minutes with a hacksaw, it's just as plain as day. For example, your technician says, "This op amp failed, all by itself, and I was just sitting there, watching it, not doing anything." But when you look inside, one of the input's lead-bond wires has blown out, evaporated, and in the usage circuit, there are only a couple $10 \text{ k}\Omega$ resistors connected to it. Well, you can't blow a lead bond with less than 300 mA. Something must have bumped against that input lead and shorted it to a source that could supply half an ampere. There are many cases where looking inside the part is very educational. When a capacitor fails, or a trim-pot, I get my hammer and pliers and cutters and hack-saw and look inside just to see how nicely it was (or wasn't) built. To see if I can spot a failure mechanism—or a bad design. I'm just curious. But sometimes I learn a lot.

Now, when I have finished my inspection, and I am still mad as hell because I have wasted a lot of time being fooled by a bad component—what do I do? I usually WIDLARIZE it, and it makes me feel a lot better. How do you WIDLARIZE something? You take it over to the anvil part of the vice, and you beat on it with a hammer, until it is all crunched down to tiny little pieces, so small that you don't even have to sweep it off the floor. It sure makes you feel better. And you know that that component will never vex you again. That's not a joke, because sometimes if you have a bad pot or a bad capacitor, and you just set it aside, a few months later you find it slipped back into your new circuit and is wasting your time again. When you WIDLARIZE something, that is not going to happen. And the late Bob Widlar is the guy who showed me how to do it.

Troubleshooting by Phone—A Tough Challenge

These days, I do quite a bit of troubleshooting by telephone. When my phone rings, I never know if a customer will be asking for simple information or submitting a routine application problem, a tough problem, or an insoluble problem. Often I can give advice just off the top of my head because I know how to fix what is wrong. At other times, I have to study for a while before I call back. Sometimes, the circuit is so complicated that I tell the customer to mail or transmit the schematic to me. On rare occasions, the situation is so hard to analyze that I tell the customer to put the circuit in a box with the schematic and a list of the symptoms and ship it to me. Or, if the guy is working just a few miles up the road, I will sometimes drop in on my way home, to look at the actual problem.

Sometimes the problem is just a misapplication. Sometimes parts have been blown out and I have to guess what situation caused the overstress. Here's an example: In

June, a manufacturer of dental equipment complained of an unacceptable failure rate on LM317 regulators. After a good deal of discussion, I asked, "Where did these failures occur?" Answer: North Dakota. "When did they start to occur?" Answer: In February. I put two and two together and realized that the climate in a dentist's office in North Dakota in February is about as dry as it can be, and is conducive to very high electrostatic potentials. The LM317 is normally safe against electrostatic discharges as high as 3 or 4 kV, but walking across a carpeted floor in North Dakota in February can generate *much* higher voltages than that. To make matters worse, the speed-control rheostat for this dental instrument was right out in the handle. The wiper and one end of the rheostat were wired directly to the LM317's ADJUST pin; the other end of the rheostat was connected to ground by way of a 1 kΩ resistor located back in the main assembly (see Figure 1.3). The speed-control rheostat was just wired up to act as a lightning rod that conducted the ESD energy right into the ADJUST pin.

The problem was easily solved by rewiring the resistor in series with the IC's ADJUST pin. By swapping the wires and connecting the rheostat wiper to ground (see Figure 1.4), much less current would take the path to the ADJUST pin and the diffused resistors on the chip would not be damaged or zapped by the current surges. Of course, adding a small capacitor from the ADJ pin to ground would have done just as well, but some customers find it easier to justify moving a component than adding one . . . .

A similar situation occurs when you get a complaint from Boston in June, "Your op amps don't meet spec for bias current." The solution is surprisingly simple: Usually a good scrub with soap and water works better than any other solvent to clean off the residual contaminants that cause leakage under humid conditions. (Fingerprints, for example . . .) Refer to Chapter 5 for notes on how a dishwasher can clean up a leaky PC board—or a leaky, dirty IC package.

When Computers Replace Troubleshooters, *Look Out*

Now, let's think—*what* needs troubleshooting? Circuits? Television receivers? Cars?[2] People? Surely doctors have a lot of troubleshooting to do—they listen to

2. If you don't think troubleshooting of cars can be entertaining, tune in *Car Talk* with Tom and Ray Magliozzi. Ask your local National Public Radio station for the broadcast time . . . GOOD STUFF!

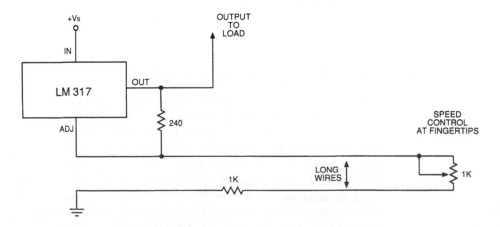

Figure 1.3. When you walk across a dry carpet and reach for the speed control, you draw an arc and most of the current from the wiper of the pot goes right into the LM317's ADJ pin.

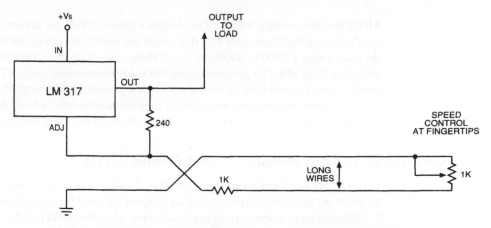

Figure 1.4. By merely swapping two wires, the ESD pulse is now sent to ground and does no harm.

symptoms and try to figure out the solution. What is the natural temptation? To let a computer do all the work! After all, a computer is quite good at listening to complaints and symptoms, asking wise questions, and proposing a wise diagnosis. Such a computer system is sometimes called an Expert System—part of the general field of Artificial Intelligence. But, I am still in favor of *genuine intelligence*. Conversely, people who rely on Artificial Intelligence are able to solve some kinds of problems, but you can never be sure if they can accommodate every kind of Genuine Stupidity as well as Artificial Stupidity. (That is the kind that is made up especially to prove that Artificial Intelligence works just great.)

I won't argue that the computer isn't a natural for this job; it will probably be cost-effective, and it won't be absent-minded. But, I am definitely nervous because if computers do all the routine work, soon there will be nobody left to do the thinking when the computer gives up and admits it is stumped. I sure hope we don't let the computers leave the smart troubleshooting people without jobs, whether the object is circuits or people.

My concern is shared by Dr. Nicholas Lembo, the author of a study on how physicians make diagnoses, which was published in the New England Journal of

Medicine. He recently told the Los Angeles Times, "With the advent of all the new technology, physicians aren't all that much interested (in bedside medicine) because they can order a $300 to $400 test to tell them something they could have found by listening." An editorial accompanying the study commented sadly: "The present trend . . . may soon leave us with a whole new generation of young physicians who have no confidence in their own ability to make worthwhile bedside diagnoses." Troubleshooting is still an art, and it is important to encourage those artists.

The Computer Is Your Helper . . . and Friend . . . ???

I read in the San Francisco *Chronicle* (Ref. 5) about a case when SAS, the Scandinavian airline, implemented an "Expert System" for its mechanics:

"Management knew something was wrong when the quality of the work started decreasing. It found the system was so highly mechanized that mechanics never questioned its judgment. So the mechanics got involved in its redesign. They made more decisions on the shop floor and used the computer to augment those decisions, increasing productivity and cutting down on errors. 'A computer can never take over everything,' said one mechanic.'Now there are greater demands on my judgment, (my job) is more interesting.'" What can I add? Just be thoughtful. Be careful about letting the computers take over.

No Problems?? No Problem . . . Just Wait . . .

Now, let's skip ahead and presume we have all the necessary tools and the right receptive attitude. What else do we need? What is the last missing ingredient? That reminds me of the little girl in Sunday School who was asked what you have to do to obtain forgiveness of sin. She shyly replied, "First you have to sin." So, to do troubleshooting, first you have to have some trouble. But, that's usually not a problem; just wait a few hours, and you'll have plenty. Murphy's Law implies that if you are not prepared for trouble, you will get a lot of it. Conversely, if you have done all your homework, you may avoid most of the possible trouble.

I've tried to give you some insights on the philosophy of how to troubleshoot. Don't believe that you can get help on a given problem from only one specific person. In any particular case, you can't predict who might provide the solution. Conversely, when your buddy is in trouble and needs help, give it a try—you could turn out to be a hero. And, even if you don't guess correctly, when you do find out what the solution is, you'll have added another tool to *your* bag of tricks.

When you have problems, try to think about the right plan to attack and nail down the problem. When you have intermittent problems—those are the nastiest types—we even have some advice for that case. (It's cleverly hidden in Chapter 12.) So, if you do your "philosophy homework," it may make life easier and better for you. You'll be able not only to solve problems, but maybe even to avoid problems. That sounds like a good idea to me!

References

1. Dostal, Jiří, *Operational Amplifiers*. Elsevier Scientific, The Netherlands, 1981; also, Elsevier Scientific, Inc., 655 Avenue of the Americas, NY, NY 10010. (212) 989-3800 ($113 in 1990)
2. Smith, John I., *Modern Operational Circuit Design*, John Wiley & Sons, New York, NY, 1971.

3. *Data Converter Handbook*, Analog Devices Corp., P.O. Box 9106, Norwood MA 02062, 1984.
4. Bulleid, H. A. V., *Master Builders of Steam*, Ian Allan Ltd., London, UK, 1963, pp. 146-147.
5. Caruso, Denise, "Technology designed by its users," The San Francisco *Examiner*, p. E-15, Sunday, March 18, 1990.

2. Choosing the Right Equipment

As discussed in Chapter 1, the most important thing you need for effective trouble-shooting is your wits. In addition to those, however, you'll normally want to have some equipment. This chapter itemizes the equipment that is necessary for most general troubleshooting tasks; some you can buy off the shelf, and some you can build yourself.

Before you begin your troubleshooting task, you should know that the equipment you use has a direct bearing on the time and effort you must spend to get the job done. Also know that the equipment you need to do a good job depends on the kind of circuit or product you are working on. For example, a DVM may be unnecessary for troubleshooting some problems in digital logic. And, the availability and accessibility of equipment may present certain obstacles. If you only have a mediocre oscilloscope and your company can't go out and buy or rent or borrow a fancy full-featured scope, then you will have to make do.

If you lack a piece of equipment, be aware that you are going into battle with inadequate tools; certain clues may take you much longer than necessary to spot. In many cases when you spent too much time finding one small problem, the time was wasted simply because you were foolish or were unaware of a particular troubleshooting technique; but, in other cases, the time was wasted because of the lack of a particular piece of equipment. It's important for you to recognize this last-mentioned situation. Learning when you're wasting time because you lack the proper equipment is part of your education as a troubleshooter.

In addition to the proper tools, you also need to have a full understanding of how both your circuit and your equipment are supposed to work. I'm sure you've seen engineers or technicians work for many fruitless hours on a problem and then, when they finally find the solution, say, "Oh, I didn't know it was supposed to work that way." You can avoid this scenario by using equipment that you are comfortable and familiar with.

The following equipment is essential for most analog-circuit troubleshooting tasks. This list can serve as a guide to both those setting up a lab and those who just want to make sure that they have everything they need—that they aren't missing any tricks.

1. A dual-trace oscilloscope. It's best to have one with a sensitivity of 1 or 2 mV/cm and a bandwidth of at least 100 MHz. Even when you are working with slow op amps, a wide-bandwidth scope is important because some transistors in "slow" applications can oscillate in the range of 80 or 160 MHz, and you should be able to see these little screams. Of course, when working with fast circuits, you may need to commandeer the lab's fastest scope to look for glitches. Sometimes a peak-to-peak automatic triggering mode is helpful and time-saving. Be sure you know how all the controls work, so you don't waste much time with setup and false-triggering problems.

2. Two or three scope probes. They should be in good condition and have suitable

hooks or points. Switchable 1×/10× probes are useful for looking at both large and very small signals. You should be aware that 1× probes only have a 16- or 20-MHz bandwidth, even when used with a 100-MHz scope. When you use 10× probes, be sure to adjust the capacitive compensation of the probe by using the square-wave calibrator per Figure 2.1. Failure to do so can be a terrible time-wasting source of trouble.

Ideally, you'll want three probes at your disposal, so that you can have one for the trigger input and one for each channel. For general-purpose troubleshooting, the probes should have a long ground wire, but for high-speed waveforms you'll need to change to a short ground wire (Figure 2.2) The shorter ground wires not only give you better frequency response and step response for your signal, but also better rejection of other noises around your circuit.

In some high-impedance circuits, even a 10× probe's capacitance, which is typically 9 to 15 pF, may be unacceptable. For these circuits, you can buy an active probe with a lower input capacitance of 1.5 to 3 pF ($395 to $1800), or you can build your own (Figure 2.3).

When you have to work with switching regulators, you should have a couple of current probes, so you can tell what those current signals are doing. Some current probes go down to DC; others are inherently AC coupled (and are much less expensive).

3. An analog-storage oscilloscope. Such a scope can be extremely useful, especially when you are searching for an intermittent or evanescent signal. The scope can trigger off an event that may occur only rarely and can store that event and the events that follow it. Some storage scopes are balky or tricky to apply, but it's often worthwhile to expend the effort to learn how to use them. Digital-storage oscilloscopes (DSOs) let you do the same type of triggering and event storage as do the analog type, and some can display events that precede the trigger. They are sampled-data systems, however, so you must be sure to apply them correctly (Ref. 1). Once you learn how to use them, though, you'll appreciate the special features they offer, such as bright CRT displays, automatic pulse-parameter measurements, and the ability to obtain plots of waveforms.

4. A digital voltmeter (DVM). Choose one with at least five digits of resolution, such as the HP3455, the HP3456, the Fluke 8810A, or the Fluke 8842A. Be sure you can lock out the autorange feature, so that the unit can achieve its highest accuracy and

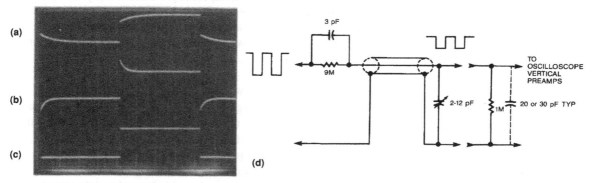

Figure 2.1. If an amplifier or a comparator is supposed to produce a square wave but the waveform looks like trace (a) or (b), how long should it take you to find the problem? No time at all! Just turn the screw that adjusts the 10× probe's compensation, so the probe's response is flat at all frequencies (c). The schematic diagram of a typical 10× oscilloscope probe is shown in (d).

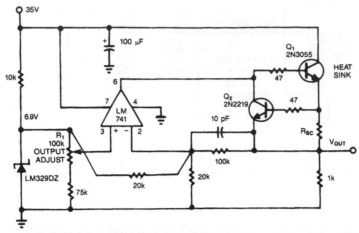

Figure 2.5. You can vary the output voltage of this DC power supply from 3 to 30 V by adjusting R_1. R_{sc} should be between 3 and 100 Ω; the short-circuit current is equal to about 20 mA + 600 mV/ R_{sc}.

don't let you continuously sweep the voltage up and down while you monitor the scope and watch for trends. In cases when the power supply's output capacitor causes problems, you may want a power supply whose output circuit, like that of an op amp, includes no output capacitor. You can buy such a supply, or you can make it with an op amp and a few transistors. The advantage of the supply shown in Figure 2.5 is that you can design it to slew fast when you want it to.

(For speed, use a quick LF356 rather than a slow LM741). Also, if a circuit latches and pulls its power supply down, the circuit won't destroy itself by discharging a big capacitor.

While we are on the topic of power, another useful troubleshooting tool is a set of batteries. You can use a stack of one, two, or four 9-V batteries, ni-cads, gel cells, or whatever is suitable and convenient. Batteries are useful as an alternate power supply for low-noise preamplifiers: If the preamp's output doesn't get quieter when the batteries are substituted for the ordinary power supply, don't blame your circuit troubles on the power supply. You can also use these batteries to power low-noise circuits, such as those sealed in a metal box, without contaminating their signals with any external noise sources.

8. A few RC substitution boxes. You can purchase the VIZ Model WC-412A, which I refer to affectionately as a "Twiddle-box" (Figure 2.6) from R & D Electronics, 1432 South Main Street, Milpitas CA 95035, (408) 262-7144. Or, inquire at VIZ, 175 Commerce Drive, Fort Washington, PA 19034, (800) 523-3696. You can set the unit in the following modes: R, C, series RC, parallel RC, open circuit, or short circuit. They are invaluable for running various "tests" that can lead to the right answer.

You may need component values beyond what the twiddle boxes offer; in our labs, we built a couple of home-brew versions (Figure 2.7). The circuit shown in Figure 2.7a provides variable low values of capacitance and is useful for fooling around with the damping of op amps and other delicate circuits. You can make your own calibration labels to mark the setting of the capacitance and resistance values. The circuit shown in Figure 2.7b provides high capacitances of various types, for testing power supplies and damping various regulators.

9. An isolation transformer. If you are working on a line-operated switching regulator, this transformer helps you avoid lethal and illegal voltages on your test setup and on

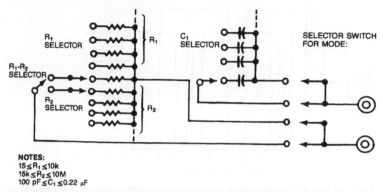

NOTES:
$15 \leq R_1 \leq 10k$
$15k \leq R_2 \leq 10M$
$100 \text{ pF} \leq C_1 \leq 0.22 \ \mu F$

Figure 2.6. This general schematic is for a commercially available RC substitution box, the VIZ Model WC-412A. The unit costs around $139 in 1991 dollars and has resistor and capacitor values in the range of 15 Ω to 10 MΩ and 100 pF to 0.2 μF, respectively. It can be configured to be an open circuit, a series RC, resistors, capacitors, a parallel RC, or a short circuit. See text for availability.

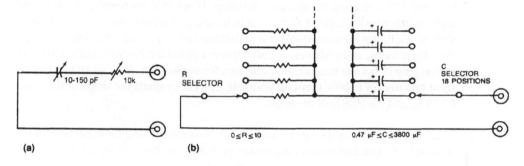

(a)   (b)

Figure 2.7. RC boxes based on these schematics extend component ranges beyond those available in off-the-shelf versions. You can house the series RC circuit in (a) in a 1 × 1 × 2-inch copper-clad box. Use the smaller plastic-film-dielectric tuning capacitors or whatever is convenient, and a small 1-turn pot. Build the circuit in (b) with tantalum or electrolytic (for values of 1 μF and higher) capacitors, but remember to be careful about their polarities and how you apply them. Also, you might consider using mylar capacitors for smaller values. Sometimes it's very valuable to compare a mylar, a tantalum, and an aluminum electrolytic capacitor of the same value! Use 18-position switches to select R and C values. And, stay away from wirewound resistors; their inductance is too high.

the body of your scope. If you have trouble obtaining an isolation transformer, you can use a pair of transformers (step-down, step-up) back-to-back (Figure 2.8). Or, if cost isn't an issue, you can use isolated probes. These probes let you display small signals that have common-mode voltages of hundreds of volts with respect to ground, and they won't require you to wear insulated gloves when adjusting your scope.

10. A variable autotransformer, often called a Variac™.[1] This instrument lets you change the line voltage and watch its effect on the circuit—a very useful trick. (Warning: A variable autotransformer is *not* normally an isolation transformer. You may need to cascade one of each, to get safe adjustable power.)

---

1. Registered TM of GENRAD Corp., Concord, MA. Variacs can be purchased from JLM Electronics, 56 Somerset St., P.O. Box 10317, West Hartford, CT 06110, (203) 233-0600.

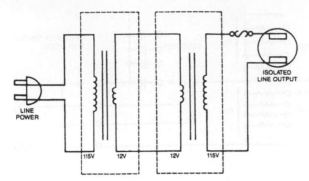

Figure 2.8. You can use this back-to-back transformer configuration to achieve line isolation similar to that of an isolation transformer.

11. A curve tracer. A curve tracer can show you that two transistors may have the same saturation voltage under a given set of conditions even though the slope of one may be quite different from the slope of the other. If one of these transistors works well and the other badly, a curve tracer can help you understand why. A curve tracer can also be useful for spotting nonlinear resistances and conductances in diodes, capacitors, light bulbs, and resistors. A curve tracer can test a battery by loading it down or recharging it. It can check semiconductors for breakdown. And, when you buy the right adapters or cobble them up yourself, you can evaluate the shape of the gain, the CMRR, and the PSRR of op amps.

12. Spare repair parts for the circuit-under-test. You should have these parts readily available, so you can swap components to make sure they still work correctly.

13. A complete supply of resistors and capacitors. You should have resistors in the range from $0.1 \, \Omega$ to $100 \, M\Omega$ and capacitors from 10 pF to $1 \, \mu F$. Also, 10, 100, and 1000 μF capacitors come in handy. Just because your circuit design doesn't include a $0.1 \, \Omega$ or a $100 \, M\Omega$ resistor doesn't mean that these values won't be helpful in troubleshooting it. Similarly, you may not have a big capacitor in your circuit; but, if the circuit suddenly stops misbehaving when you put a 3800 μF capacitor across the power supply, you've seen a quick and dramatic demonstration that power-supply wobbles have a lot to do with the circuit's problems. Also, several feet of plastic-insulated solid wire (telephone wire) often come in handy. A few inches of this type of twisted-pair wire makes an excellent variable capacitor, sometimes called a "gimmick." Gimmicks are cheap and easy to vary by simply winding or unwinding them. Their capacitance is approximately one picofarad per inch.

14. Schematic diagrams. It's a good idea to have several copies of the schematic of the circuit-under-test. Mark up one copy with the normal voltages, currents, and waveforms to serve as a reference point. Use the others to record notes and waveform sketches that relate to the specific circuit-under-test. You'll also need a schematic of any homemade test circuit you plan to use. Sometimes, measurements made with your homemade test equipment may not agree with measurements made by purchased test equipment. The results from each tester may not really be "wrong": They might differ because of some design feature, such as signal filtering. If you have all the schematics for your test equipment, you can more easily explain these incompatibilities. And, finally, the data sheets and schematics of any ICs used in your circuit will also come in handy.

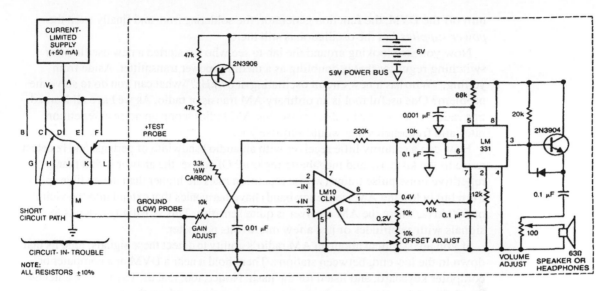

Figure 2.9. You can use this short-circuit detector to find PC board shorts. Simply slide the test probe along the various busses and listen for changes in pitch.

15. Access to any engineering or production test equipment, if possible. Use this equipment to be sure that when you fix one part of the circuit, you aren't adversely affecting another part. Other pieces of equipment and testers also fall under the category of specialized test equipment; their usefulness will depend on your circuit. Three examples are a short-circuit-detector circuit, an AM transistor radio, and a grid-dip meter.

   A short-circuit-detector circuit. This tool comes in handy when you have to repair a lot of large PC boards: It can help you find a short circuit between the ground bus and the power or signal busses. It's true that a sensitive DVM can also perform this function, but a short-circuit detector is much faster and more efficient. Also, this circuit turns itself off and draws no current when the probe is not connected. In the short-circuit-detector circuit shown in Figure 2.9, the LM10 op amp amplifies the voltage drop and feeds it to the LM331 voltage-to-frequency converter, which you set up to emit its highest pitch when $V_{in} = 0$ mV. When using this circuit, use a 50- to 100-mA current-limited power supply. To calibrate the circuit, first ground the detector's two probes and trim the OFFSET ADJUST for a high pitch. Then, move the positive test probe to V_S at A and trim the GAIN ADJUST for a low pitch. Figure 2.9 illustrates a case in which one of the five major power supply busses of the circuit-in-trouble has a solder short to ground.

   To find the exact location of the short, you simply slide the positive input probe along the busses. In this example, if you slide it from A toward B or D, the pitch won't change because there is no change in voltage at these points—no current flowing along those busses. But, if you slide the probe along the path from A to C or from K to M, the pitch will change because the voltage drop is changing along those paths. It's an easy and natural technique to learn to follow the shifting frequency signals.

   An AM radio. What do you do when trouble is everywhere? A typical scenario starts out like this: You make a minor improvement on a linear circuit, and when you fire it up you notice a terrible oscillation riding on the circuit's output. You check everything about the circuit, but the oscillation remains. In fact, the oscillation is riding on the output, the inputs, on several internal nodes, and even on ground. You turn off the DVM, the function generator, the soldering iron, and finally even the *power supplies*, but the oscillation is still there.

Now you start looking around the lab to see who has started a new oscillator or switching regulator that is doubling as a medium-power transmitter. Aside from yelling, "Who has a new circuit oscillating at 87 kHz?" what can you do to solve the problem? One useful tool is an ordinary AM transistor radio. As we have all learned, FM radios reject many kinds of noise, but AM radios scoop up noise at repetition rates and frequencies that would surprise you.

How can a crummy little receiver with an audio bandwidth of perhaps 5 kHz detect noise in the kilohertz and megahertz regions? Of course, the answer is that many repetitive noise-pulse trains (whose repetition rates are higher than the audible spectrum but below the AM frequency band) have harmonics that extend into the vicinity of 600 kHz, where the AM receiver is quite sensitive. This sensitivity extends to signals with amplitudes of just a few microvolts per meter.

If you are skeptical about an AM radio's ability to detect these signals, tune its dial down to the low end, between stations. Then, hold it near a DVM or a computer or computer keyboard, and listen for the hash. Notice, too, that the ferrite stick antenna has definite directional sensitivity, so you can estimate where the noise is coming from by using either the null mode or pointing the antenna to get the strongest signal. So, the humble AM radio may be able to help you as you hike around the lab and smile pleasantly at your comrades until you find the culprit whose new switching regulator isn't working quite right but which he neglected to turn off when he went out to get a cup of coffee.

The grid-dip meter.   On other occasions, the frequency and repetition rate of the noise are so high that an AM receiver won't be helpful in detecting the problem. What's the tool to use then? Back in the early days of radio, engineers found that if you ran a vacuum-tube oscillator and immersed it in a field of high-power oscillations at a comparable frequency, the tube's grid current would shift or dip when the frequencies matched. This tool became known as the "grid-dip meter." I can't say that I am an expert in the theory or usage of the grid-dip meter, but I do recall being impressed in the early days of monolithic ICs: A particular linear circuit was oscillating at 98 MHz, and the grid-dip meter could tickle the apparent rectified output error as I tuned the frequency dial back and forth.

That was 25 years ago, and, of course, Heathkit[1] has discontinued their old Grid Dip and Tunnel Dip meters in favor of a more modern design. The new one, simply dubbed HD-1250 "Dip Meter," uses transistors and tetrode FETs. At the bargain price of $89, every lab should have one. They'll help you ferret out the source of nasty oscillations as high as 250 MHz. The literature that comes with the HD-1250 dip-meter kit also lists several troubleshooting tips.

When grid-dip meters first became popular, the fastest oscilloscope you could buy had a bandwidth of only a few dozen megahertz. These days, it is possible to buy a scope with a bandwidth of many hundreds of megahertz, so there are fewer occasions when you might need a grid-dip meter. Still, there are times when it is exactly the right tool. For example, you can use its oscillator to activate passive tuned circuits and detect their modes of resonance. Also, in a small company where you can't afford to shell out the many thousands of dollars for a fast scope, the dip meter is an inexpensive alternative.

16. A few working circuits, if available. By comparing a bad unit to a good one, you can

---

1. Heath Company, Benton Harbor, Michigan, 49022; (1-800-253-0570).

often identify problems. You can also use the good circuits to make sure that your specialized test equipment is working properly.

17. A sturdy, broad workbench. It should be equipped with a ground plane of metal that you can easily connect to the power ground. The purpose of this ground plane is to keep RF, 60-Hz, and all other noise from coupling into the circuit. Place insulating cardboard between the bench and the circuit-under-test, so that nothing tends to short to ground. Another way to prevent noise from interfering with the circuit is to use a broad sheet of single-sided copper-clad board. Placed copper-side down and with a ground wire soldered to the copper, it provides an alternate ground plane. To prevent electrostatic-discharge (ESD) damage to CMOS circuits, you'll need a wrist strap to ground your body through 1 MΩ.

18. Safety equipment. When working on medium- or high-power circuits that might explode with considerable power in the case of a fault condition, you should be wearing safety goggles or glasses with safety lenses. Keep a fire extinguisher nearby, too.

19. A suitable hot soldering iron. If you have to solder or unsolder heavy busses from broad PC-board traces, use a big-enough iron or gun. For small and delicate traces around ICs, a small tip is essential. And, be sure that the iron is hot enough. An easy way to delaminate a trace or pad, whether you want to or not, is to heat it for too long a time, which might happen if your iron weren't big enough or hot enough. (The old Heathkit warnings not to use a hot iron became obsolete along with the germanium transistor.) In some cases, a grounded soldering iron is required; in others, a portable (ungrounded or rechargeable) soldering iron is ideal. Make sure you know whether your iron is grounded or floating.

20. Tools for removing solder, such as solder wick or a solder sucker. You should be comfortable with whatever tools you are using; a well-practiced technique is sometimes critical for getting good results. If you are working on static-sensitive components, an antistatic solder-sucker is less likely to generate high voltages due to internal friction than is an ordinary solder-sucker. I have been cautioned that a large solder-sucker may cause problems when working on narrow PC traces; in that case, solder wick may be the better choice.

21. Hand tools. Among the tools you'll need are sharp diagonal nippers, suitable pliers, screwdrivers, large cutters, wrenches, wire strippers, and a jack knife or Exacto™ knife.

22. Signal leads, connectors, cables, BNC adapters, wires, clip leads, ball hooks, and alligator clips—as needed. Scrimping and chintzing in this area can waste lots of time: shaky leads can fall off or short out.

23. Freeze mist and a hair dryer. The freeze mist available in aerosol cans lets you quickly cool individual components. A hair dryer lets you warm up a whole circuit. You'll want to know the dryer's output air temperature because that's the temperature to which you'll be heating the components.

    NOTE: Ideally we should not use cooling sprays based on chlorofluorocarbons (CFCs), which are detrimental to the environment. I have a few cans that some people would say I shouldn't use. But what else should I do—send the can to the

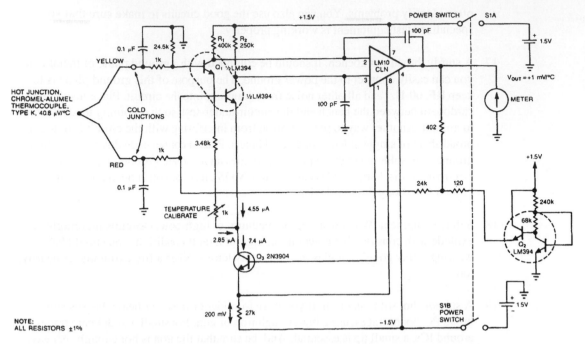

Figure 2.10. This thermocouple amplifier has inherent cold-junction compensation because of the two halves of Q_1, which run at a 1.6:1 current ratio. Their V_{BE}s are mismatched by 12 mV + 40.8 μV/°C. This mismatch exactly cancels out the 40.8 μV/°C of the cold junction. For best results, you should use four 100 kΩ resistors in series for R_1 and two 100 kΩ resistors in series with two 100 kΩ resistors in parallel for R_2—all resistors of the same type, from the same manufacturer. Q_2 and its surrounding components implement a correction for very cold temperatures and are not necessary for thermocouple temperatures above 0 °C. Credit to Mineo Yamatake for his elegant circuit design.

dump? Then it will soon enter the atmosphere, without doing anybody any good. I will continue to use up any sprays with CFC-based propellants that I already have, but when it is time to buy more, I'll buy environmentally safe ones.

24. A magnifying glass or hand lens. These devices are useful for inspecting boards, wires, and components for cracks, flaws, hairline solder shorts, and cold-soldered joints.

25. An incandescent lamp or flashlight. You should be able to see clearly what you are doing, and bright lights also help you to inspect boards and components.

26. A thermocouple-based thermometer. If your thermometer is floating and battery powered, you can connect the thermocouple to any point in the circuit and measure the correct temperature with virtually no electrical or thermal effect on the circuit. Figure 2.10 shows a thermocouple amplifier with designed-in cold-junction compensation.
    Some people have suggested that an LM35 temperature-sensor IC (Figure 2.11) is a simple way to measure temperature, and so it is. But, if you touch or solder an LM35 in its TO-46 package to a resistor or a device in a TO-5 or TO-3 case, the LM35 will increase the thermal mass and its leads will conduct heat away from the device whose temperature you are trying to measure. Thus, your measurements will be less accurate than if you had used a tiny thermocouple with small wires.

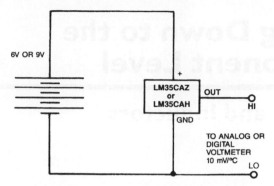

Figure 2.11. The LM35CAZ is a good, simple, convenient general-purpose temperature sensor. But beware of using it to measure the temperature of very small objects or in the case of extreme temperature gradients; it would then give you less accurate readings than a tiny thermocouple with small wires.

27. Little filters in neat metal boxes, to facilitate getting a good signal-to-noise ratio when you want to feed a signal to a scope. They should be set up with switch-selectable cut-off frequencies, and neat connectors. If in your business you need sharp roll-offs, well, you can roll your own. Maybe even with op-amps and batteries. *You* figure out what you need. Usually I just need a couple simple Rs and Cs, with an alligator clip to select the right ones.

28. Line adapters—those 3-wire-to-2-wire adapters for your 3-prong power cords. You need several of them. You only need them because too many scopes and function generators have their ground tied to the line-cord's neutral. You need some of these to avoid ground-loops. You also need a few spares because your buddies will steal yours. For that matter, keep a few spare cube taps. When they rewired our benches a few years ago, the electricians tried to give us five outlets per bench. I stamped my feet and insisted on ten per bench, and that's just barely enough, most of the time.

You've come to the end of my list of essential equipment for ordinary analog-circuit troubleshooting. Depending on your circuit, you may not need all these items; and, of course, the list did not include a multitude of other equipment that you may find useful. Logic analyzers, impedance analyzers, spectrum analyzers, programmable current pumps, capacitance meters and testers, and pulse generators can all ease various troubleshooting tasks.

Each of you will have your own idea of what is essential and what is unnecessary for your special case, and I would be delighted to get feedback on this subject. You can write to me at the address in the Acknowledgments section of this book.

References

1. Collins, Jack, and David White, "Time-domain analysis of aliasing helps to alleviate DSO errors," *EDN*, September 15, 1988, p. 207.

3. Getting Down to the Component Level

Resistors and Inductors

In earlier chapters, we've covered the philosophy of troubleshooting analog circuits, and the tools and equipment you need to do so. But if you're working on a circuit and are not aware of what can cause component failure, finding the root of your problem could be difficult. Hence, this chapter covers resistors, fuses, inductors, and transformers; their possible modes of failure; and the unsuspected problems that may occur if you use the wrong type of component. (Capacitors are waiting in the next chapter. Kind of a shame to segregate them from the resistors . . . .)

Troubleshooting circuits often boils down to finding problems in passive components. These problems can range from improper component selection in the design phase to damaged components that hurt the circuit's performance. Resistors, inductors, and transformers can each be a source of trouble.

Resistors are certainly the most basic passive component, and, barring any extreme or obscure situations, you won't usually run into problems caused by the resistors themselves. I don't mean to say that you'll never see any resistor problems, but most of them will be due to the way you use and abuse and mis-specify resistors. In other cases, some other part of the circuit may be causing damage to a resistor, and the failure of the resistor is just a symptom of a larger problem.

You may eventually have to track down a wide variety of problems involving resistors to achieve a working design. Some will seem obvious. For example, your circuit needs a 10 kΩ resistor. The technician reaches into the drawer for one and instead gets a 1 kΩ resistor, which then mistakenly gets inserted into your board. This example illustrates the most common source of resistor trouble in our lab. Consequently, I ask my technicians and assemblers to install resistors so that their values are easy to read. And any time I find a 1 k resistor where a 10 k is supposed to be, I check to see how many more 1 ks are in the 10 k drawer. Often there are quite a few!

Sometimes a resistor gets mismarked; sometimes a resistor's value shifts due to aging, overheating, or temperature cycling. Recently, we found a batch of "1%" metal-film resistors whose values had increased by 20 to 900% after just a few dozen cycles of −55 to +125 °C. As it turned out, our QC department had okayed only certain resistors to be used in burn-in boards, and these particular resistors had not been approved. The QC people, too, had spotted this failure mode.

Resistor Characteristics Can Vary Widely

You should be familiar with the different resistor types in order to select the most appropriate type for your application; the most common types and some of their characteristics are summarized in Table 3.1. A component type that's good for one application can be disastrous in another. For example, I often see an engineer specify a carbon-composition resistor in a case where stability and low TC (Temperature Coefficient) are required. Sometimes it was just a bad choice, and a conversion to a stable metal-film resistor (such as an RN55D or RN60C) with a TC of 50 or 100 ppm/°C max considerably improves accuracy and stability. In other cases, the engineer says, "No, I tried a metal-film resistor there, but, when I put in the carbon resistor, the overall TC was improved." In this case, the engineer was relying on the carbon-composition resistor to have a consistent TC which must compensate for some other TC problem. I have found that you can't rely on consistent TC with the carbon-composition type, and I do not recommend them in applications where precision and stability are required—even if you do see some TC improvement in your circuit.

However, carbon-composition resistors do have their place. I was recently reviewing a military specification that spelled out the necessary equipment for the ESD (electrostatic discharge) testing of circuits. An accurate 1500 Ω resistor was required for use as the series resistor during discharge of the high-voltage capacitor. In this case, you would assume that a metal-film resistor would be suitable; however, a metal-film resistor is made by cutting a spiral into the film on the resistor's ceramic core (Figure 3.1a). Under severe overvoltage conditions, the spiral gaps can break down and cause the resistor to pass a lot more current than Ohm's Law predicts—the resistor will start to destroy itself. Therefore, the spec should have called for the use of a carbon-composition resistor, whose resistive element is a large chunk of resistive material (Figure 3.1b). This resistor can handle large overloads for a short time without any such flash-over. Even when you are applying a 200% to 400% overload for just a short time, the nonuniform heating of the spiraled section of a metal-film resistor can cause the resistor to become unreliable. You can also get around this problem by using a series connection of metal-film resistors. If you put fifteen 100 Ω, 1/4-W metal-film resistors in series, each individual resistor would not see overvoltage or excessive power.

Carbon-film resistors are now quite inexpensive and have become the most common type of resistor around most labs. Their main drawback is that they are very similar in appearance to metal-film resistors and have some similar characteristics: Carbon-film resistors have 1% tolerances, are normally manufactured with spiral cuts, and have the same kind of voltage-overload limitations as metal-film types. But, carbon-film resistors have much higher TCs—500 to 800 ppm/°C. It's easy to erroneously insert a drifty carbon-film resistor for the intended metal-film type. Don't confuse the two.

Precision-film resistors, on the other hand, are available with greatly improved accuracy and TC. Compared to ordinary RN55D and RN55C resistors with TCs of

Table 3.1. Typical Resistor Characteristics

Resistor Type	Range* (Ω)	TC (±PPM/°C)	Parasitic Effects	Cost
Composition	1–22M	High	Low	Low
Metal Film	10–1M	Low	Medium	Medium
Carbon Film	10–10M	Medium	Medium	Medium
Wirewound, Precision	1–1M	Low	High	High
Wirewound, Power	0.01–100k	Medium	High	Medium
Thin-Film	25–100k	Low	Low	Medium
Thick-Film	10–1M	Low	Low	Medium
Diffused	20–50k	High	High	Low

*Range may vary by manufacturer.

"voltage coefficient." This last term refers to the nonlinearities in Ohm's Law that occur when there is a large voltage drop across a resistor; the effect is most common in resistors with large values and small values—ones built with high densities.

Therefore, when you drive the reference input to a D/A converter, you should be aware that the R_{in} will only shift 1–3% over the entire temperature range. However, there may still be a broad tolerance, as it is not easy to keep tight tolerances on the "sheet rho," or resistivity, during the IC's production. For example, a typical D/A converter's R_{in} specification is 15 kΩ ±33%. These film resistors have even better tracking TC than diffused resistors, often better than 1 ppm/°C.

In addition to the TC, you might also be concerned with the shunt capacitance of a resistor. Recently (back in Chapter 2), I was trying to build a high-impedance probe with low shunt capacitance. I wanted to put a number of 2.5 MΩ resistors in series to make 10 MΩ. I measured the shunt capacitance of several resistors with our lab's impedance bridge. A single Allen-Bradley carbon-composition resistor had a 0.3 pF capacitance, so the effective capacitance of four in series would be down near 0.08 pF—not bad (Figure 3.3). Then I measured a Beyschlag carbon-film resistor. Its capacitance was slightly lower, 0.26 pF. The capacitance of a Dale RN60D was 0.08 pF; the capacitance of four in series would be almost unmeasurable.

It would be an improper generalization to state that certain resistor types (films) always have less shunt capacitance than others. However, the main point is that if you need a resistance with low shunt capacitance, you can connect lower-value resistors in series; and if you evaluate several different manufacturers' resistors, you may find a pleasant surprise.

Variable Resistors and Pots

As with the fixed resistors discussed so far, there are many kinds and types of variable resistors, such as trimming potentiometers, potentiometers, and rheostats. These resistors are made with many different resistive elements, such as carbon, cermet, conductive plastic, and wire. As with fixed resistors, be careful of inexpensive carbon resistors, which may have such poor TC that the manufacturer avoids any mention of it on the data sheet. These carbon resistors would have a poor TC when used as a rheostat but might have a good TC when used as a variable voltage divider or a potentiometer. Recently I ran an old operational amplifier where the offset trim pot had a range of 100 mV. Yet for 4 hours in a row, the amplifier's offset held better than ±10 μV. That's an amazing ±0.01% stability for a carbon composition pot! On the other hand, some of the cermet resistors have many excellent characteristics but are not recommended for applications that involve many hundreds of wiper cycles. For example, a cermet resistor would be inappropriate for a volume control on a radio.

The major problem area for variable resistors is their resolution, or "settability." Some variable resistors claim to have "infinite" resolution; but, if you apply 2 V across a variable resistor's ends and try to trim the wiper voltage to any or every millivolt in between, you may find that there are some voltage levels you can't achieve. So much for "infinite resolution." As a rule of thumb, a good pot can usually be set to a resolution of 0.1%, or every 2 mV in the previous case. Thus, counting on a settability of 0.2% is conservative.

Good settability includes not only being able to set the wiper to any desired position but also having it stay there. But, I still see people advertising multi-turn pots with the claimed advantage of superior settability. The next time you need a pot with superior settability, evaluate a multi-turn pot and a single-turn pot. Set each one to the desired value, tap the pots with a pencil, and tell me which one stays put. I normally expect a multi-turn pot, whether it has a linear or circular layout, to be a factor

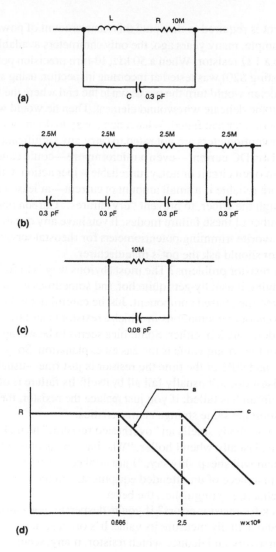

Figure 3.3. You can reduce the capacitance of a single resistor (a) by using several resistors in series as shown in (b) (assume the inductance is negligible). This series resistor configuration has one-fourth of the single resistor's capacitance (c) and extends the resistor's frequency response as shown in (d).

of 2 to 4 worse than a single-turn pot because the mechanical layout of a single-turn pot is more stable and balanced. Does anyone know of an example in which the multi-turn pot is better? A full year after this statement was originally published, nobody has tried to contradict me, although people who sell multi-turn pots still brag in the vaguest possible terms about "infinite resolution"—bleah!

Don't Exceed Your Pot's I and V Ratings

How do variable resistors fail? If you put a constant voltage between the wiper and one end and turn the resistance way down, you will exceed the maximum wiper current rating and soon damage or destroy the wiper contact. Note that the power rating of most variable resistors is based on the assumption that the power dissipation is uniformly distributed over the entire element. If half of the element is required to dissipate the device's rated power, the pot may last for a short while. However, if a quarter of the element is required to dissipate this same amount of power, the pot will fail quickly. For example, many years ago, the only ohmmeters available might put as much as 50 mA into a 1 Ω resistor. When a 50 kΩ, 10-turn precision potentiometer (think of an item costing $20) was tested at incoming inspection using such an ohmmeter, the test technician would turn the pot down to the end where the 50 mA was sufficient to burn out the delicate wirewound element. Then he would write in his report that the potentiometer had failed. What a dumb way to do incoming inspection!

Some trimming potentiometers are not rated to carry any significant DC current through the wiper. This DC current—even a microampere—could cause electromigration, leading to an open circuit or noisy, unreliable wiper action. Other trim-pots are alleged to be more reliable if a small amount of current—at least a microampere DC—*is* drawn through the wiper, to prevent "dry failures." Carbon pots are not likely to be degraded by either of these failure modes. If you have any questions about the suitability of your favorite trimming potentiometers for rheostat service, you or your components engineer should ask the pot's manufacturer.

How do you spot resistor problems? The most obvious way is to follow your nose. When a resistor is dying it usually gets quite hot, and sometimes the strong smell of phenolic leads right to the abused component. Just be careful not to burn your fingers. You may also encounter situations in which a resistor hasn't truly failed but doesn't seem to be doing its job, either. Something seems to be wrong with the circuit, and a resistor of the wrong value is the easiest explanation. So, you measure the resistor in question, and 90% of the time the resistor is just fine—usually the trouble is elsewhere. A resistor doesn't usually fail all by itself. Its failure is often a symptom that a transistor or circuit has failed; if you just replace the resistor, the new one will also burn out or exhibit the same strange characteristics.

Around our lab, if anybody smells an "overheated resistor," he makes sure that we understand what it is. Usually when I holler, "Who has a resistor overheated??," an engineer or technician will sheepishly say, "I just cooked my circuit . . . ." But sometimes it is a failure in a piece of un-attended equipment, and the sooner we can turn off its power or fetch a fire extinguisher, the better.

How do you check for resistor errors? If you're desperate, you can disconnect one end of the resistor and actually measure its value. It's often easier to just measure the I $\times$ R drops in the network and deduce which resistor, if any, seems to be of the wrong value. If one resistor is suspected of being temperature-sensitive, you can heat it with a soldering iron or cool it with freeze mist as you monitor its effect. In some solid-state circuits, the signals are currents, so it's not easy to probe the circuit with a voltmeter. In this case, you may have to make implicit measurements to decide if a resistor is the problem. Also, remember that a sneak path of current can often cause the same effect as a bad resistor.

When you are trying to make precision measurements of resistors, you should be aware that even the best ohmmeters—even the ones with 4-wire connections and lots of digits on the DVM—do not have as good accuracy or resolution as you can get by forcing a current through a stable reference resistor R_{REF} and then through the R_X and comparing the voltages. This is especially true for low resistor values. See Figure 3.4. You also have complete control over the amount of current flowing through the R_X.

Watch Out for Damaged Components

Damaged resistors can also be the source of trouble. A resistor that's cracked can be noisy or intermittent. When resistors are overheated with excess power, such as 2 or 3 W in a 1/4-W resistor, they tend to fail "open"—they may crack apart, but they

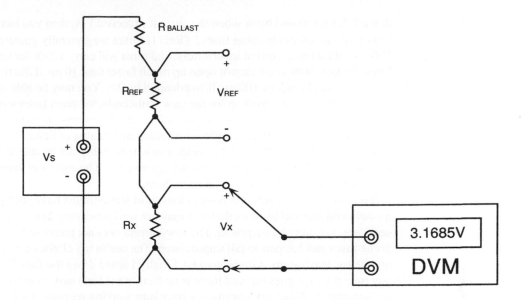

Figure 3.4. If you use a good voltmeter to measure V_{ref} and V_X and take the ratio, you can resolve the R_X a lot better than in the OHMS mode.

don't go to low ohms or to a short circuit. The accuracy or stability of a high-value resistor (10^8 to 10^{12} Ω) can be badly degraded if dirt or fingerprints touch its body. Careful handling and cleaning are important for these high-value resistors and high-impedance circuits.

One problem that occurs with all resistors is related to the Seebeck effect: the production of an EMF in a circuit composed of two dissimilar metals when their two junctions are at different temperatures. In precision circuits, you should avoid thermal gradients that could cause a large temperature difference across a critical resistor. For example, don't stand a precision resistor on end, as in an old transistor radio—if it has any dissipation, it might get a lot hotter on one end than the other. Many precision wirewound and film resistors have low Seebeck coefficients in the range 0.3 to 1.5 μV/°C. But avoid tin oxide resistors, which can have a thermocouple effect as large as 100 μV/°C. If you are going to specify a resistor for a critical application where thermocouple errors could degrade circuit performance, check with the manufacturer.

So, you ought to know that resistors can present challenging troubleshooting problems. Rather than re-inventing the wheel every time, try to learn from people with experience.

When Is a Resistor Not Just a Resistor?

When it's a fuse. Obviously, when a low-value resistor is fed too much current and fails "open," that is sometimes a useful function, and the multi-million-dollar fuse industry thus serves to protect us from trouble. But the fuses themselves can cause a little trouble. They don't always blow exactly when we wish they would. As Ian Sinclair put it in his book *Passive Components—a User's Guide*,[2] "If you thought

2. Mr. Sinclair's book has a lot of good information in it, about all kinds of passive components, and I thoroughly recommend it—see Ref. 1 for more information.

that a 1 A fuse would blow when the current exceeded 1 A, then you have not been heavily involved in choosing fuses." (Ref. 1) Fuses are generally guaranteed to carry 100% of their rated current indefinitely, and most will carry 120% for several hours. Even the fast-blow ones cannot open up much faster than 10 ms if overloaded by 10× their rated load, or 100 ms if overloaded by 2×. You may be able to get faster response than that if you shop for the new semiconductor-rated fuses with very fast blow time. If somebody in your organization—a components engineer or an old-timer—can help you find the right information in a catalog of fuses, he can save you a lot of time. Without that kind of help, you will probably not be able to find a catalog from a fuse maker, or to figure it out when you get it. The curves of various ratings are a little obscure until you get used to them.

You may not use fuses much—modern solid-state circuits have such good current-limiters and thermal limiters that you may not see fuses every day. So when you do see fuses, you may be surprised. The low-current ones act pretty soft—resistive. Some fuses just happen to fail unprovoked. The one in my clothes-dryer fails every 3 or 4 years, leaving my wife perplexed. Finally I wrote down the list of symptoms, so any time the fuse goes out and there is no heat, we at least save time by recognizing the symptoms. When my microwave oven quit working recently, I was a little concerned because the label on its back said, "No user-serviceable components inside." When I opened it up, there was a fuse clip with a blown fuse. After shopping unsuccessfully at several electrical supply houses, I finally went in to a Radio Shack. They had them and, I realized, that was the *first* place I should have gone. I replaced the fuse and turned on the power—would the fuse blow for a good reason, or had the old fuse just fatigued out? The new fuse has held for several months, so it was just a fatigue failure.

Most fuses are fully rated for 115 or 230 VAC, but not more than 32 V of DC. That's because the alternating current flow gives time for an arc to be extinguished, which would not happen with DC. So for high-voltage DC, the answers aren't so simple. Some circuit breakers are rated for as much as 65 VDC, but often that's not enough. There is a CD Series that is good up to 125 VDC, and a larger GJ Series that is rated up to 150 V, available from Heinemann.[3]

Another approach in circuits with rectified power is to put the sensing coil in the DC circuit, but connect the breaker into the AC circuit. That's no help if you just have a 120-V battery supply.

These days, high-powered MOSFETs can be used to make such a good high-voltage high-current switch that you can build your own fast-turn-off switch, activated by over-current—an electronic equivalent of a fuse. I built one of these—and it didn't work very well, the first time. It blew out the FET. Twice. I haven't really given up on it, and when I get some time after I put this book to bed, I'll go back and get it running. When I get it running, I'll publish it somewhere where you can all see it: The equivalent of a solid-state fuse that can handle as much as 200 V of DC.

Meanwhile, when you need some fuse protection on a DC power supply, just put a fuse in the secondary of the transformer, so it sees AC current flow and AC voltage, rather than DC.

Inductors and Transformers Aren't So Simple

Inductors and transformers are more complicated than resistors—nonlinearity is rife. Their cores come in many different shapes and sizes, from toroids to pot cores and from rods to stacks of laminations. Core materials range from air to iron to any of the

3. Heinemann Electric, P. O. Box 6800, Lawrenceville, NJ 08648. (609) 882-4800.

ferrites. I am not going to presume to tell you how to design an inductor or transformer or how to design circuits that use them, but I will discuss the kinds of trouble
you can have with these components. For example, you can have a good core material; but, if there is an air gap in the core and you don't carefully control the gap's
width, the energy storage and the inductance of the component can vary wildly. If
someone has substituted a core of the wrong material, you may have trouble spotting
the change; an inductance meter or an impedance bridge can help. But even with one
of those tools, you're not home free.

For most inductors and transformers with cores composed of ferromagnetic materials, you had better make sure that the test conditions—the AC voltage and the frequency that the measuring instrument applies to the device under test—closely approximate those the component will see in your real-life application. If you fail to
take such precautions, your inductance measurements stand a good chance of seriously misleading you and making your troubleshooting task much more frustrating.
The phenomena you are likely to run into as a result of incorrect test conditions include saturation, which can make the inductance look too low, and core loss, which
can lower the Q of an inductor. For transformers, make sure you understand which of
the inductances in the device's equivalent circuit you are measuring.

Equivalent Circuits Demystify Transformers

You can represent a transformer with a turns ratio of N as a "T" network (Figure 3.5a).
N equals N_1/N_2, where N_2 is the number of secondary turns and N_1 is the number of
primary turns. However, if you plan to make measurements on transformers, it's
helpful to keep the equivalent circuit shown in Figure 3.5b in mind. For example, the
inductance you measure between terminals A and B is quite large if you leave terminals C and D open, but the measured inductance will be quite small if you short terminals C and D together. In the first case, you are measuring the mutual inductance
plus the leakage inductance of the primary. But because the leakage inductance is normally much, much smaller than the mutual inductance, you are measuring the leakage inductance of the primary plus the reflected secondary leakage in the second case.

When you work with inductors or transformers, you have to think in terms of current: In any transformer or inductor, flux is directly proportional to the current, and
resistive losses are directly proportional to the current squared. Therefore, be sure to
have several current probes, so you can observe what the current waveforms are
doing. After all, some of the weirdest, ugliest, and most nonideal waveforms you'll
see are the waveforms associated with inductors. (Especially in a switch-mode
regulator . . .)

In the absence of an instrument designed to measure inductance, parallel the inductor with a known capacitance to create a parallel resonant circuit. If you use a
high-impedance source to apply a current pulse to this circuit, you can determine the
inductor's value from the resonant frequency and the capacitance: $f = 1/(2 \pi \sqrt{LC})$.
If you look at the inductor's waveform on a scope, you can compare it to the waveform you get with a known good inductor. This technique is also good for spotting a
shorted turn, which reduces inductance nearly to zero. The L meter and the similar Q
meter can help you ensure that good inductors haven't been damaged by saturation.

Incredible as it may sound, you can permanently damage an inductor by saturating
it. Some ferrite toroids achieve their particular magnetic properties by means of operation at a particular point on the material's magnetization curve. Saturating the core
can move the operating point and drastically change the core's magnetic properties.
The likelihood of your being able to return the material to its original operating point
is small to nonexistent. In other cases, as a result of applying excessive current, the

core temperature increases to a point where the core's magnetic properties change irreversibly. Regardless of the mechanism that caused the damage, you may have to do as I once did—package the inductors with a strongly worded tag to demand that nobody test them at Incoming Inspection.

Bob Widlar had a good solution to that. He would instruct the Incoming Inspection Technician to count the number of leads. Don't *measure* anything, just *count* the number of leads. If they follow that instruction, they probably won't wreck the transformer.

If you choose too small a wire size for your windings, the wire losses will be excessive. You can measure the winding resistance with an ohmmeter, or you can measure the wire's thickness. But if the number of turns is wrong, you can best spot the error with an L meter—remember that $L \propto N^2$. Be careful when using an ohmmeter to make measurements on transformers and inductors—some ohmmeters put out so many milliamps that they are likely to saturate the component you are trying to measure and at least temporarily alter its characteristics. Select an ohmmeter which puts out only a small amount of current.

Protect Transistors from Voltage Kick

There is one trouble you can have with an inductor or relay coil that will not do any harm to the magnetic device, but will leave a trail of death and destruction among its associated components: When you use a transistor to draw a lot of current through an inductor and then turn the transistor off, the "kick" from the inductor can generate a

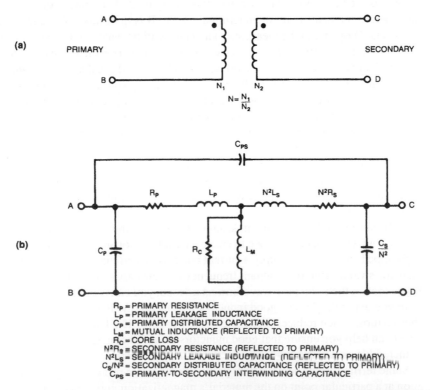

R_P = PRIMARY RESISTANCE
L_P = PRIMARY LEAKAGE INDUCTANCE
C_P = PRIMARY DISTRIBUTED CAPACITANCE
L_M = MUTUAL INDUCTANCE (REFLECTED TO PRIMARY)
R_C = CORE LOSS
N^2R_S = SECONDARY RESISTANCE (REFLECTED TO PRIMARY)
N^2L_S = SECONDARY LEAKAGE INDUCTANCE (REFLECTED TO PRIMARY)
C_S/N^2 = SECONDARY DISTRIBUTED CAPACITANCE (REFLECTED TO PRIMARY)
C_{PS} = PRIMARY-TO-SECONDARY INTERWINDING CAPACITANCE

Figure 3.5. (a) In most instances, you can represent a transformer by its turns ratio. (b) If you are measuring the characteristics of a transformer, you should keep its equivalent circuit in mind. Considering the effect of each component will help you understand the results of your measurements.

voltage high enough to damage or destroy almost any transistor. You can avoid this problem by connecting a suitable snubber, such as a diode, an RC network, a zener, or a combination of these components, across the inductor to soak up the energy. The use of a snubber is an obvious precaution, yet every year I see a relay driver with no clamp to protect the transistor. The transistor may survive for a while, but not for long.

The tiniest inductors are called beads. They are about the same size and shape as beads worn as jewelry, they are available in various types of ferrite material, and they have room for only one or two or four turns of wire. Beads are commonly used in the base or emitter of a fast transistor to help keep it from oscillating. A bead not only acts inductive but also acts lossy at high frequencies, thus damping out ringing. In general, the choice of a bead is an empirical, seat-of-the-pants decision, but designers who have a lot of experience in this area make good guesses. This topic is one that I have not seen treated (except perhaps one sentence at a time) in any book or magazine. You'll just have to get a box of ferrite beads and experiment and fool around.

Transformers usually are susceptible to the same problems as inductors. In addition, the turns ratio may be wrong, or the winding polarity might be incorrect. And, if your wire-handling skills are sloppy, you might have poor isolation from one winding to another. Most ferrite materials are insulators, but some are conductive. So, if you've designed a toroidal transformer whose primary and secondary windings are on opposite sides of the toroid and you scrape off the core's insulating coating, you could lose your primary-secondary insulation. If the insulating coating isn't good enough, you might need to wrap tape over the core.

Fortunately, it's easy to establish comparisons between a known-good transformer and a questionable one. If you apply the same input to the primaries of both transformers, you can easily tell if the secondaries are matched, wound incorrectly, or connected backwards. If you're nervous about applying full line voltage to measure the voltages on a transformer, don't worry—you can drive the primary with a few volts of signal from a function generator (preferably in series with a resistor and/or a capacitor, to prevent saturation and overload) and still see what the various windings are doing.

Two general problems can afflict power transformers. The first occurs when you have large filter capacitors and a big high-efficiency power transformer. When you turn the line power switch on, the in-rush current occasionally blows the fuse. You might install a larger value of fuse, but then you must check to make sure that the fuse is not too high to offer protection. As an alternative, you could specify the transformer to have a little more impedance in the secondary: Use smaller wire for the windings or put a small resistor in series with the secondary.

Another approach, often used in TV sets, is to install a small negative-TC thermistor in the line power's path. The thermistor starts out with a nominal impedance, so the surge currents are finite. But then the thermistor quickly heats up, and its resistance drops to a negligible value. Thus, the efficiency of the circuit is quite good after a brief interval. If the circuit is a switch-mode power supply, the control IC should start up in a "soft-start" mode. In this mode, the IC makes sure the switcher won't draw any extreme currents in an attempt to charge up the output capacitors too quickly. However, you must use caution when you apply thermistors for in-rush current limiting: Beware of removing the input power and then re-applying it before the thermistors have had a chance to cool. A hot thermistor has low resistance and will fail to limit the current; thus, you are again likely to blow a fuse—or a rectifier.

The second general problem with a line transformer occurs when you have a small output filter capacitor. In our old LM317 and LM350 data sheets, we used to show typical applications for battery chargers with just a 10 μF filter. Our premise was that when the transformer's secondary voltage dropped every 8 ms, there was no harm in

letting the regulator saturate. That premise was correct, but we began to see occasional failed regulators that blew up when we turned the power on.

After extensive investigations, we found the problem in the transformer: If the line
power switch was turned off at exactly the wrong time of the cycle, the flux in the
transformer's steel core could be stored at a high level. Then, if the line power switch
was reconnected at exactly the wrong time in the cycle, the flux in the transformer
would continue to build up until the transformer saturated and produced a voltage
spike of 70–90 V on its secondary. This spike was enough to damage and destroy the
regulator. The solution was to install a filter capacitor of at least 1000 μF, instead of
just 10 μF. This change cut the failure rate from about 0.25% to near zero.

Another problem occurred when the LM317 was used as a battery charger. When
the charger output was shorted to ground, the LM317 started drawing a lot of current.
But, the transformer's inductance kept supplying more and more current until the
LM317 went into current limit and could not draw any more current. At this point,
the transformer's secondary voltage popped up to a very high voltage and destroyed
the LM317. The addition of the 1000 μF snubber also solved this problem.

Inductors, Like Resistors, Can Overheat

How do you spot a bad inductor or transformer? I have already discussed several
mechanisms that can cause the inductance or Q of an inductor to be inferior to that of
a normal part. And, as with a resistor, you can smell an inductor that is severely overheating. Overheating can be caused by a faulty core, a shorted turn, incorrect wire
gauge, or anything else that causes losses to increase. An open winding is easy to
spot with an ohmmeter, as is a short from a primary to a secondary. If the pattern of
winding has been changed from one transformer to another, you may not see it unless
you test the components in a circuit that approximates the actual application.
However, you may also be able to see such a discrepancy if you apply a fast pulse to
the two transformers. Changes in winding pattern—even clockwise vs. counter-
clockwise—have been observed to cause significant changes in transformer performance and reliability.

Tightly-coupled windings, both bifilar and twisted pairs, have much better magnetic coupling and less leakage inductance than do well-separated primary and secondary windings. As the magnetic coupling improves, the capacitance between windings increases—but high capacitance between windings is often an undesirable effect
in a transformer. An experienced transformer designer weighs all the tradeoffs and
knows many design tricks—for example, the use of special pi windings and Litz
wire. Mostly, you should know that these special techniques are powerful; if you ask
the transformer designers the right questions, they can do amazing tricks.

I recently read about an engineer who designed an elegant shield made of mu-
metal. However, the shield was difficult to install, so the technician had to tap on it
with a hammer. When the engineer operated the circuit, the shielding seemed nonexistent—as if the shield were made of cardboard. After a lot of studying, the engineer
realized that the mu-metal—which costs about $2 per 15 square inches, the same as a
$2 bill—had been turned into perfectly worthless material by the pounding and ham-
mering. In retrospect, the engineer had to admit that the mu-metal, when purchased,
was prominently labelled with a caution against folding, bending, or hammering. So
remember, in any area of electronics, there are problems with inductors and magnetic
materials that can give you gray hair.

Consider the Effects of Magnetic Fields

One problem recently illustrated the foibles of inductor design: Our applications engineers had designed several DC/DC converters to run off 5 V and to put out various voltages, such as +15 V and −15 V DC. One engineer built his converter using the least expensive components, including a 16-cent, 300 μH inductor wound on a ferrite rod. Another engineer built the same basic circuit but used a toroidal inductor that cost almost a dollar. Each engineer did a full evaluation of his converter; both designs worked well. Then the engineers swapped breadboards with each other. The data on the toroid-equipped converter was quite repeatable. But, they couldn't obtain repeatable measurements on the cheaper version. After several hours of poking and fiddling, the engineers realized that the rod-shaped inductor radiated so much flux into the adjacent area that all measurements of AC voltage and current were affected. With the toroid, the flux was nicely contained inside the core, and there were no problems making measurements. The engineers concluded that they could tell you how to build the cheapest possible converter, but any nearby circuit would be subject to such large magnetic fields that the converter might be useless.

When I am building a complicated precision test box, I don't even try to build the power supply in the main box because I know that the magnetic fields from even the best power transformer will preclude low-noise measurements, and the heat from the transformer and regulators will degrade the instrument's accuracy. Instead, I build a separate power-supply box on the end of a 3-ft cable; the heat and magnetic flux are properly banished far away from my precision circuits.

Reference

1. *Passive Components—A User's Guide*, Ian Sinclair, Heinemann Newnes, Halley Court, London, England. 1990, p. 225. Order from Butterworth-Heinemann, 80 Montvale Avenue, Stoneham, Mass. 02180.

though this leakage is usually quite low, nobody wants to have to measure it in production, nor to guarantee it for the lifetime of the component.

Wound-film and stacked-film capacitors cover wide ranges, from small signal-coupling capacitors to large high-power filters. The different dielectrics are their most interesting ingredients. Often a designer installs a polyester capacitor (technically, polyethylene terephthalate, often called Mylar—a trademark of E.I. DuPont de Nemours and Co.) and wonders why something in the circuit is drifting 2 or 3% as the circuit warms up. What's drifting is probably the polyester capacitor; its TC of 600 to 900 ppm/°C is 10 times as high as that of a metal-film resistor.

If you give up on polyester and go to polystyrene, polypropylene, or Teflon, (also a trademark of DuPont) the TC gets better—about –120 ppm/°C. Polystyrene and polypropylene have low leakage and good dielectric absorption—almost as good as Teflon's, which is the best (Ref. 1). But Teflon is quite expensive and rather larger in package size than the other types. Be careful with polystyrene; its maximum temperature is +85 °C, so you might damage it during ordinary wave-soldering unless you take special precautions to keep the capacitors from over-heating. Polycarbonate, polysulfone, and polyphenylene have good TCs of about +100 ppm/°C, and their names have enough syllables that they sound as if they should be pretty good, but actually they have inferior soakage. Glass and porcelain are dielectrics that sound like they ought to have some really fancy characteristics, and excellent dielectric absorption. But they don't, not very good at all. Many years ago, wound-film capacitors were made with oil-impregnated paper, but you won't see them unless you are working on ancient radios. They were pretty crummy, just adequate for audio coupling on low-fidelity radios.

Foiled Again!

Now let's discuss the difference between a polyester foil capacitor and a metallized polyester capacitor. The foil capacitor is made of alternating layers of film and foil, where both the delicate film and the metal foil are just a couple of tenths of a milli-inch thick. This construction makes a good capacitor at a nominal price and in a nominal size. The metallized-film capacitor is made with only a very thin film of polyester—with the metal deposited on the polyester in a *very* thin layer. This construction leads to an even smaller size for a given capacitance and voltage rating, but the deposited metal is so thin that its current-carrying capacity is much less than that of the metal in the foil capacitor. This offers advantages and disadvantages. If a pinhole short develops in this metallized-polyester capacitor's plastic film, the metal layer in the vicinity of the pinhole will briefly carry such a high current density that it will vaporize like a fuse and "clear" the short.

For many years, metallized polyester capacitors were popular in vacuum-tube television sets because they were small and cheap. These metallized capacitors would recover from pinhole flaws not just once but several times. However, at low voltages, the energy stored in the capacitors would often prove insufficient to clear a fault. Thus, the capacitors' reliability at low voltages was often markedly worse than it was at their rated voltage. You could safely use a cheap, compact, metallized-polyester capacitor in a 100-V TV circuit but not in a 2-V circuit. Fortunately, there are now classes of metallized-polycarbonate, metallized-polyester, and metallized-polypropylene capacitors that are reliable and highly suitable for use at both low and high voltages. I was reading one of these data sheets the other day, and it said that at low voltages, any pin-hole fault is cleared by means of oxidation of the ultra-thin metal film.

When the old metallized-polyester capacitors began to become unreliable in a TV set, the "clearing" of the shorts would make the signals very noisy. Likewise, when used as audio coupling capacitors, "dry" tantalum capacitors would sometimes make a lot of noise as they "cleared" their leaky spots. These parts have therefore become unpopular for audio coupling. Similarly, you might use an electrolytic capacitor with a small reverse voltage—perhaps 0.5 V—with no harm or problems. BUT a friend told me of the time he was using an electrolytic capacitor as an audio-coupling capacitor with 2 V of reverse bias. Because of the reverse bias, it was producing *all sorts of* low-frequency noise and jitter. So, excess noise is often a clue that something is going wrong—perhaps it is trying to tell you about a misapplication, or a part installed backwards.

Extended Foil Offers Extensive Advantages

Another aspect of the film capacitor is whether or not it uses "extended-foil" construction. The leads of many inexpensive wound-foil capacitors are merely connected to the tip ends of the long strip of metal foil. However, in an extended-foil capacitor, the foils extend out on each side to form a direct low-resistance, low-inductance path to the leads.

This construction is well suited for capacitors that must provide low ESR (equivalent series resistance) in applications such as high-frequency filters. Then if you substituted a capacitor without extended foil, the filter's performance would be drastically degraded.

So there are several methods of construction and several dielectrics that are important considerations for most capacitor applications. If an aggressive purchasing agent wants to do some substituting to improve cost or availability, the components engineer or design engineer may have to do a lot of work to make sure that the substitution won't cause problems. If a substitution is made, the replacement part is a good place to start looking for trouble. A capacitor with higher-than-planned-for ESR can cause a feedback loop to oscillate—for example, when a capacitor without extended-foil construction is substituted for one with such construction. Substitution of capacitors with higher ESR than the designer intended can also cause filters to fail to prop-

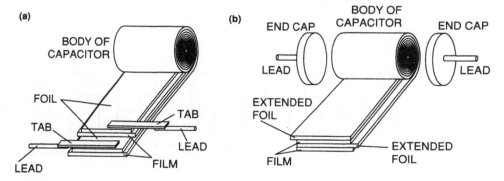

Figure 4.1 (a) When the tabs connect to one end of a long foil, some elements of the capacitor will be 10 or 20 feet away from the leads. The series Rs and Ls are poor. This construction was adequate for low-fidelity audio circuits, but is uncommon these days. (b) When the exposed edges of the extended foils are crimped together, no element of the capacitor is more than an inch or two from the leads and connections. Most film capacitors are made with extended foils these days.

erly attenuate ripple. Another consequence of excessive ESR is the overheating and failing of capacitors—capacitors may be passive components, but they are not trivial.

Not only does extended-foil construction lower a capacitor's ESR, it also lowers the component's inductance. As a friend, Martin Giles, pointed out, after reading a draft of my troubleshooting text, "Pease, you understand things really well if they are at DC or just a little bit faster than DC." I replied, "Well, that's true, but what's your point?" His point was that in RF circuits, and many other kinds of fast circuits, you should use capacitors and other components dressed closely together, so that the inductance is small and well controlled. He is absolutely right—the layout of a high-speed, fast-settling or a high-frequency circuit greatly affects its performance. Capacitors for such circuits must be compact and not have long leads. Ceramic and silvered-mica capacitors are often used for that reason.

Every year, billions of ceramic capacitors find their way into electronic products of all kinds. There are basically three classes of these parts: the "high-K" and "stable-K" types and the C0G or NP0 types.

The high-K types, such as those with a "Z5U" characteristic, give you a lot of capacitance in a small space—for example, 10^6 pF in a 0.3-in. square that is 0.15-in. thick. That's the good news. The bad news is that the capacitance of parts with this Z5U characteristic drops 20% below the room-temperature value at 0 and 55 °C; it drops 60% below the room-temperature value at –25 and +90 °C. Also, the dielectric has a poor dissipation factor, mediocre leakage, and a mediocre voltage coefficient of capacitance. Still, none of these drawbacks prevents capacitors of this type from being used as bypass capacitors across the supply terminals of virtually every digital IC in the whole world. That's a lot of capacitors!

These ceramic capacitors have a feature that is both an advantage and a drawback—a typical ESR of 0.1 Ω or lower. So, when a digital IC tries to draw a 50-mA surge of current for a couple of nanoseconds, the low ESR is a good feature: It helps to prevent spikes on the power-supply bus. To get good bypassing and low inductance you must, of course, install the ceramic capacitors with minimum lead length. However, when you have 10 ICs in a row and 10 ceramic bypass capacitors, you've got a long LC resonator (Figure 4.2) with the power-supply bus acting as a low-loss inductor between each pair of bypass capacitors. When repetitive pulses excite this resonator, ringing of rather large amplitude can build up and cause an excessively noisy power-supply bus. This can be especially troublesome if the signal

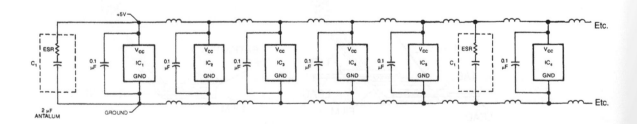

Figure 4.2. Low ESR in a decoupling capacitor is a two-edged sword. Though a capacitor with low ESR stabilizes the supply bus when the ICs draw short-duration current spikes, the low dissipation factor encourages ringing by allowing the decoupling capacitance to resonate with the bus inductance. One good cure is to place electrolytic capacitors, such as C_1, across the bus. C_1's ESR of approximately 1 Ω damps the ringing.

rep rates are close to the resonant frequency of the LC network! And remember that the Z5U capacitors have a poor TC, so that as the circuit warms up, it really is likely that there will be a temperature where the ringing frequency moves up to be a multiple of the clock frequency.

The standard solution is to add 2 μF of tantalum electrolytic bypass capacitors or 20 μF of aluminum electrolytic capacitors for every three to five ICs (unless you can prove that they are unnecessary). That's a good rule of thumb. The ESR of the electrolytic capacitors, typically 1 Ω, is essential to damp out the ringing. Some people say that this ESR is too high to do any good in a bypass capacitor—but they do not understand the problem. I have read a few ads in which some capacitor manufacturers claim that their ceramic bypass capacitors are so good—have such low series resistance—that ringing is no longer a problem. I find the claims hard to believe. I invite your comments.

ESR, Friend or Foe?

Specifically, some capacitor manufacturers claim that the series resistance, R_S, is so low that you won't have a problem with ringing. But low R_S would seem to exacerbate the ringing problem. Conversely, I've heard that one capacitor manufacturer is proposing to market ceramic capacitors whose series R_S has a lower limit—a few ohms—to help damp out any ringing. I'm going to have to look into that. But if you have bypass capacitors with a very low R_S, you can lower the Q of the resonator you have inadvertently constructed around them by adding a resistor of 2.7 to 4.7 Ω in series with some of the capacitors. Adding resistance in series with bypass capacitors might seem a bit silly, but it's a very useful trick.

High-K ceramic capacitors also can exhibit piezoelectric effects: When you put a good amount of AC voltage across them, they can hum audibly; and if you rattle or vibrate them, they can kick out charge or voltage. (Other types can do the same thing, but high-K types are worse.) Be careful when using these capacitors in a high-vibration environment.

The capacitance of stable-K capacitors, such as X7R, typically decreases by less than 15% from the room-temperature value over the –55 to +125 °C range. These capacitors are general-purpose devices and are usually available in the 100 to 10,000 pF range; in the larger packages, you can get as much as 300,000 pF. However, you can buy a 10,000 pF capacitor in either a high-K or a stable-K type; and you can't be sure of the kind you're getting unless you check the catalog and the part number. Or, measure the capacitance as you heat or cool it.

The last type of ceramic capacitor was originally called "NP0" for Negative-Positive-Zero, and is now usually called "C0G." Everybody calls them "COG," (C-oh-G) but it really is C-zero-G. I've seen the EIA document (Ref. 2). The C0G / NP0 capacitors have a really high-grade low-K dielectric with a guaranteed TC of less than ±30 ppm/°C. Their dissipation factor, dielectric absorption, and long-term stability are not quite as good as those of Teflon capacitors but are comparable to those of other good precision-film capacitors. And the TC is better than almost anything you can buy. So, if you want to make a sample-and-hold circuit usable over the military temperature range, you'll find that C0G capacitors are more compact and less expensive then Teflon parts. Many, but not all, ceramic capacitors smaller than 100 pF are made with the C0G characteristic. You can get a 22,000 pF C0G capacitor in a 0.3-in.-square package, if you're willing to pay a steep price.

About every year or so, a customer calls me about a drift problem: His V/F converter has a poor TC, even though he said that he had put in a C0G 0.01 μF capacitor as the main timer. Troubleshooting by phone—it's always a wonderful challenge. I ask him, "This C0G-ceramic 0.01 µF capacitor . . . is it . . . as big as your little fingernail?" He says, "Oh, no, it's a lot smaller than that." I reply, "Well, that's too small; it can't be a C0G." Problem solved. Actually, there are *some* small C0G 0.01µF capacitors, but they are pretty uncommon unless you order them specially.

One observed failure mode for ceramic capacitors can arise when the capacitor's leads are attached to the dielectric with ordinary, low-temperature solder. When the capacitor goes through a wave-solder machine, the lead may become disconnected from the capacitor. If this problem occurs, you'll have to switch to capacitors from a manufacturer that uses high-temperature solder.

Don't Forget Silvered Mica

Silvered-mica capacitors have many features similar to C0G capacitors. They have low ESR and a TC of 0 to +100 ppm/°C. They can also work at temperatures above 200 °C if assembled with high-temperature solder. Unfortunately, they have poor soakage characteristics—unexpectedly bad dielectric absorption.

A major problem with silver-mica capacitors is their marking. The silver-mica capacitors in old radios had completely inscrutable markings—six color dots. Some of the new ones have such odd codes that even if the marking on the capacitor hasn't rubbed off, you can never be sure whether "10C00" means 10, 100, or 1000 pF. You really need to use some kind of capacitance meter. Similarly, in the old days, some ceramic capacitors were marked in an inscrutable way. I remember two little capacitors both marked "15K." One was a 15 pF capacitor with a "K" characteristic, and the other was a 15,000 pF capacitor—yet they were both the same size and had the same marking.

I must also mention that, in the past, you could buy a pretty good capacitor that had never been tested for its capacitance. About 99% of the time, they were excellent, reliable capacitors. But once in a while, some of the capacitors came through with a capacitance value completely different from the marked value. One time I saw a

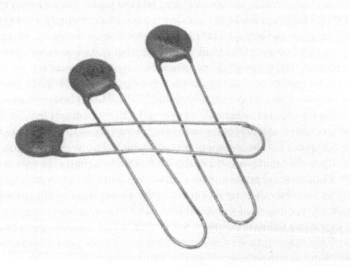

Figure 4.3.　If you saw a capacitor that looked like this, you'd know the manufacturer hadn't tested it before shipping, right?? (Photo by Steve Allen.)

whole box of "capacitors" in which the two leads were still made of one loop of wire that had not been snipped apart.

Obviously, the manufacturer wasn't interested in testing and measuring these capacitors before sending them out the door! So, if you are buying capacitors to a 1% AQL (Acceptance Quality Level) and not 0.1% or 0.01%, you should be aware that some low-priced parts have not even been sample tested.

Variable Capacitors May Have Finite Rotational Lives

Variable capacitors are usually made of low-K material with characteristics similar to those of C0G capacitors. Their electrical performance is excellent. The dielectric doesn't cause much trouble, but the metal sliding contacts or electrodes are, on some models, very thin; after only a small number of rotations—hundreds or even dozens—the metal may wear out and fail to connect to the capacitance.

In general, capacitors are very reliable components; and, if you don't fry them with heat or zap them mercilessly, the small-signal ones will last forever and the electrolytic ones will last for many years. (Old oil-filled capacitors aren't quite that reliable and have probably been replaced already—at least they should have been replaced.) The only way you can have an unreliable capacitor is to use a type that is unsuitable for the task. And that's the engineer's fault, not the capacitor's fault. Still, some troubleshooting may be required; and if you recognize the clues that distinguish different types of capacitors, you've taken a step in the right direction.

First, Try Adding a Second

What procedures are best for troubleshooting capacitors? I use two basic procedures, the first of which is the add-it-on approach. Most circuits are not hopelessly critical about capacitor values, as long as the capacitors' values are large enough. So, if there is a 0.01 μF capacitor that I suspect of not doing its job, I just slap another 0.01 μF capacitor across it. If the ripple or the capacitor's effect changes by a factor of two, the original capacitor was probably doing its job and something else must be causing the problem. But if I observe little or no change or a change of a factor of three, five, or ten, I suspect that capacitor's value was not what it was supposed to be. THEN I pull the capacitor out and measure it. Of course, the capacitor substitution boxes I mentioned in the section on test equipment, part 8 of Chapter 2, can be valuable here; they let me fool around with different values. But in critical circuits, the lead length of the wires going to the substitution box can cause crosstalk, oscillation, or noise pickup; so I may have to just "touch in" a single capacitor to a circuit.

Suppose, for example, that I have a polyester coupling capacitor that seems to be adding a big, slow "long tail" to my circuit's response. I don't expect the performance with the polyester capacitor to be perfect, but a tail like this one is ridiculous! (Note: when a capacitor's voltage is supposed to settle, but there is actually a "long tail," that is just another way of saying that the capacitor has poor dielectric absorbtion or "soakage." It's the same thing with different aspects.) So, I lift up one end of the polyester capacitor and install a polypropylene unit of the same value. I expect the new capacitor's characteristics to be a lot better than those of the old capacitor. If the tail gets a lot smaller, either my plan to use polyester was not a good one or this particular polyester capacitor is much worse than usual. It's time to check. But usually, I'd expect to find that the polypropylene capacitor *doesn't* make the circuit perform much better than the polyester capacitor did, and I'd conclude that something else must be causing the problem.

5. Preventing Material and Assembly Problems

PC Boards and Connectors, Relays and Switches

In addition to your choice of components, the materials you use to assemble your circuit will have an impact on how well it performs. This chapter covers what you need to know to solve the occasional problems caused by PC boards, solder, connectors, wire, and cable. Also covered is PC-board layout—a poor layout can cause more than occasional problems; it can determine completely how well your circuit works.

Some of the topics discussed in this troubleshooting series so far may have seemed obvious. But far too often it is this "obvious" information that engineers overlook, and it is this information that can make troubleshooting so much easier. So, be careful not to overlook the obvious. Don't assume that your PC-board materials or layout don't matter or that wire characteristics don't differ; you'll find that PC boards, connectors, wire, and cable can cause problems when you least expect it.

First of all, the use of the term "printed-circuit board" is a misnomer; these days, almost every board is an etched-circuit board. But I'll continue to use the abbreviation "PC board" because it's a familiar term. There are six basic troubles you can get into with PC boards:

- The board is made of the wrong material.

- The quality of the vendor's board is so bad that there are opens or shorts in the board or, worse yet, intermittent connections in the plated-through holes.

- The foil starts to peel off the board because of mistreatment.

- You were so concerned with cost that you neglected to specify a layer of solder mask; you ended up with a board full of solder shorts.

- The surface of the board is leaky or contaminated.

- Your circuit layout is such that signals leak and crosstalk to each other, or controlled-impedance lines are interfering, thus causing reflections and ringing.

Avoid PC-Board Problems at the Outset

The fixes for these problems, and ways to avoid them in the first place, are fairly straightforward.

These days, the G10 and G11 fiberglass-epoxy materials for PC boards are quite good and reasonably priced. Trying to use cheaper phenolic or "fishplate" is not economical in most cases. Conversely, a special high-temperature material or an exotic material or flexible substrate may be justifiable. If you don't have an expert on these materials, the PC-board maker or the manufacturer of the substrate material can usually provide some useful advice. (See Table 5.1 for a comparison of PC-board materials.)

In some RF applications, phenolic material has advantages over glass epoxy; it has a lower dielectric constant and superior dimensional stability. And for ultra-broadband oscilloscope probes, some types of glass epoxy have a definite disadvantage due to mediocre dielectric absorption, especially if the epoxy has not been properly cured.

As for quality, there is almost never an excuse for buying your boards from a vendor whose products are of unknown quality. "Low cost" would be one poor alibi; "Can't get acceptable delivery time from our normal vendor" would be another. One time, to meet a rush contract, we had to build circuits on boards made in our own lab facilities. I had never had trouble before using these boards for prototyping, so I was surprised when I began the troubleshooting and found that an apparently good board occasionally had a short between two busses.

Close inspection with a magnifying glass showed a hairline short about 3 mils wide, which was caused by a hair that fell onto the artwork. You would never ask a printed-circuit foil that narrow to carry 20 mA, but this narrow short would carry 200 mA before blowing out. Similarly, we found hairline opens: The ground bus was broken in two or three places by a tiny 4-mil gap, just barely visible to the naked eye. Of course these "opens" were caused by the image of a hair, during a negative process. After several hours of fiddling around, opening shorts, and shorting opens, we vowed not to be caught by such poor workmanship again.

As for the third problem, don't let ham-fisted engineers or technicians beat up a good PC board with the overenthusiastic or misguided application of a soldering iron. That's sure to lift the foil. Use an iron that's hot enough so you can get *in* and get *out*. If the iron's not hot enough and it's taking too long, *that's* when the foil will lift . . .

Solder mask, the subject of the fourth problem, is almost always worthwhile, as many people have learned. Without it, the admirable tendency of solder to bridge things together, which is wonderful in most instances, becomes disastrous.

Table 5.1 PC-Board Laminate Materials

Type	Manufacturer	Dielectric Constant (at 1 MHz)	Dissipation Factor (at 1 MHz)	Volume Resistivity (MΩ-cm)	Surface Resistivity (MΩ)	Maximum Temperature (°C)	Comments
XXXPC	Generic	4.1	0.032	5×10^6	5×10^4	+125	Low cost, paper-based phenolic; poor mechanical strength
CEM-1	Generic	4.5	0.025	1×10^8	5×10^7	+130	Standard, economical
CEM-3	Generic	4.7	0.020	1×10^8	5×10^7	+130	Similar to CEM-1 but can be punched
G-10	Generic	4.75	0.023	5×10^8	4×10^8	+130	Comparable to CEM-3
F4486	Oak	3.5	0.02	1×10^9	3×10^6	N/A	Flexible
FR-4	Generic	4.9	0.018	1×10^8	5×10^7	+130	Similar to CEM-3 but fire retardant per UL-94-V-0
GT-522	Keene	2.5	0.0010	1×10^7	1×10^7	+260	Teflon is good for high temp, high speed
GX-527	Keene	2.5	0.0019 (at 10 GHz)	1×10^7	1×10^7	+260	Characterized for high frequency
HI-3003	Technoply	4.5	0.020	3×10^7	5×10^6	+250 (10,000 hours)	Polyimide
3003-quartz	Technoply	3.6	0.004	5×10^9	8×10^7	+250 (10,000 hours)	Comparable to CEM-3

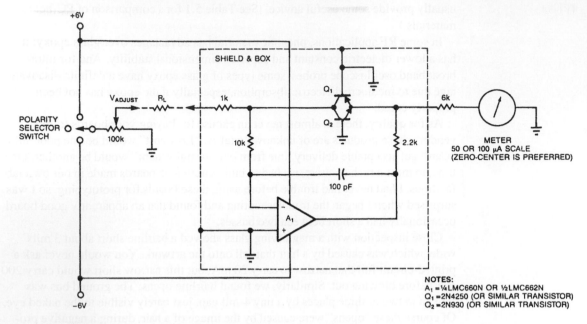

Figure 5.1. You can use this circuit to test your board for leakage current. The transistors are connected across the op amp's feedback path so that they act like a current-to-voltage detector, with a wide-range logarithmic characteristic.

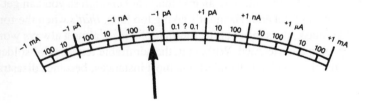

Figure 5.2. You can calibrate the logarithmic-current-meter circuit of Figure 5.1 to sense currents between −1 mA and −1 pA and between +1 pA and +1 mA.

Leakage Can Be a Problem

When a PC board comes from its manufacturer, it is usually very clean and exhibits high impedance. Sometimes a board starts out leaky, but normally a board doesn't begin to leak until you solder it or wash it with a contaminated solvent—the fifth problem.

Looking for leakages

When you have a leaky board or a slightly less-than-infinite-impedance connector or insulator, how do you test for leakage? You can't just slap a DVM on it, because even on the highest range, (for example, 20 MΩ) the display will just read Over-range. That's no help if you want to read 2000 MΩ or 20,000 or 200,000 MΩ—or even higher. Some DVMs or digital multimeters have a scale for microSiemens (conductance) which will let you resolve as high as 100 MΩ. But this scale does not normally have more resolution than that.

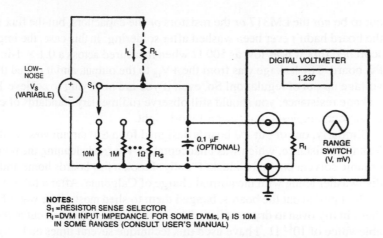

Figure 5.3. The DVM approach is an alternative to the approach illustrated in Figure 5.1 for testing leakage. You can calculate the leakage current from Ohm's Law: $V_S = I_L \times R_S$, or $I_L = V_S/R_S$.

There are basically two ways to measure leakage current. The approach I have used for many years is to connect a couple of transistors as a wide-range, logarithmic, current-to-voltage detector across the feedback path of a low-bias-current op amp. These days, I'm not using vacuum tubes—I have graduated to an LMC660, as shown in the circuit of Figure 5.1. I calibrated the meter with a hand-drawn scale to sense currents ranging from +1 pA to +1 mA and –1 pA to –1 mA (Figure 5.2). As long as the air conditioning doesn't break down, I know my calibration won't drift much more than 10 or 20%, which is adequate to tell me which decade of current I am working in. (The V_{BE} does have some temperature sensitivity, but not enough to bother this circuit very much.) Because these transistors are, of course, quite non-linear as current-to-voltage sensors, you do have to shield the summingpoint away from AC noises (60 Hz, 120 Hz, 1 MHz, etc.) to prevent rectification and false readings. So the whole test circuit and the unknown impedance are best located in a shallow metal box, grounded, with an optional metal cover.

It's true that the DVM approach shown in Figure 5.3 has a little more accuracy and perhaps more resolution; but it too is easily fooled by noise, and the digital readout doesn't show trends well. And, if you want to cover a wide range of currents, you have to switch in different resistors *or* wait for the DVM to autorange, which is not my idea of fun. On the other hand, you can find a DVM almost anywhere, so this approach is easy to implement.

In either case, if you put 15 V across 1,000,000 MΩ and measure 15 pA, that is at least 50,000 × higher resolution than most meters that can only measure up to 20 MΩ. Whichever detector you use, apply a reasonable voltage across the unknown impedance and see where the leakage gets interesting. This method can also be used for diodes and transistor junctions. The op-amp circuit is not especially recommended for measuring the leakage of large-value capacitors; neither is the DVM approach because of the slow charging of the large capacitance, and because of soakage, or dielectric absorption, effects. But, if you're desperate and start out with a low value of R_{sense}, you can eventually get some approximate measurements.

Recently, a customer had a problem with a simple basic design using an LM317 regulator in which the circuit's impedances were fairly low—just a few hundred ohms. (The same basic circuit as in chapter 14, Figure 14.3.) After just a few minutes of operation, the output of the LM317 would start drifting badly. The cause turned out to be *not* the LM317 *or* the resistors *or* the capacitors, but the flux build-up where the board hadn't ever been washed after soldering. In this case, the impedance of the scorched flux was as low as 500 Ω when measured across a 0.1 $\times$ 1-in. area of the PC board. The leakage was from the $+V_{in}$ to the output, and it pulled the output voltage up out of regulation! So, even if you are not trying to achieve 10^{12} Ω of leakage resistance, you should still observe rudimentary standards of cleanliness, or even your simplest circuits won't work right.

Similarly, one of our PC boards designed for a S/H circuit was yielding 10^{11} Ω of leakage resistance, which was unacceptable. We tried cleaning the board with every organic solvent but had no luck. Finally, I took a few boards home and set them in the dishwasher along with the normal charge of Calgonite. After a full wash-and-rinse cycle, I pulled out the boards, banged them to shed most of the water beads, and set them in my oven to dry at 160 °F. The next day, they checked out at the more acceptable value of 10^{13} Ω . I have used this technique several times on leaky PC boards and sockets, and it works surprisingly well. It can work when alcohol, TCE, and organic solvents are not helping at all.

After you get your board clean and dry, you'll want to keep it that way. For this purpose, you may want to use a coating such as the urethane, acrylic, or epoxy-types, sprays or dips. Humiseal[1] is the pioneering name, and they have a broad catalog of different types for various production needs. In a similar vein, the guys up at Essex Junction, VT told me about some varnish made by John Armitage Co.[2] which is a rather thick, heavy high-impedance coating. It takes a while to dry, but it's pretty durable and I like it. When I was building some little 1/3-ounce modules that some scientists were going to carry up to the top of Mt. Everest, I chose a couple well-baked coats of the "Armitage" to keep the modules clean and dry; it's much lighter than potting in epoxy, which is important when a guy has to carry a scientific package on his back up to 29,000 feet.

Of course, with any of these coatings, it is not trivial to cut in and repair the circuit or change components. So your choice of a durable coating should be tempered by the awareness of how much fun it is to go in and remove the coating and do your repairs.

When I was at Philbrick, we potted most of our products in epoxy, and it gave good reliability and security. If you get a good circuit in there, well potted in epoxy, it has an excellent chance to survive forever, with no moisture getting in, and with everything held isothermal, protected from shock and physical abuse. Of course, if somebody abuses the circuit electrically and damages it, it's substantially impossible to get inside to repair it. You may have to drill down to the PC board, just to do some troubleshooting. It's fun and a challenge to have to delve in and troubleshoot a potted circuit. Sometimes the potting material adds extra stresses to components: Squeezing the resistors and capacitors can change their values, and pouring epoxy around a circuit can add significant capacitance, too. If you pot a circuit that hasn't been baked and dried out thoroughly, the moisture may get sealed inside the potted module. Epoxy can cover up a multitude of sins, but there is no substitute for good workmanship and good engineering. Potting a piece of junk usually leads only to well-potted junk.

Make sure the designers who lay out your PC board keep a list of rules to avoid troubles. For example, if your circuit has a high-impedance point and you suspect leakage might be a problem, don't run that high-impedance trace beside a power-

1. Humiseal Div. of Chase Corp, 26-60 Brooklyn-Queens Expressway, Woodside, NY 11377. (718) 932-0800.

2. Armitage MM-00941 Clear Brushing Alkyd Varnish; John Armitage & Company, 1259 Route 46, Parsippany, NJ 07054. (201) 402-9000.

supply foil—guard it with a stripe of ground foil or "guard foil" between the two. A dozen times I have heard an engineer say, "The resistivity of this glass-epoxy material is 10^{14} Ω-cm, so you can't possibly expect to have a resistance of 10^{12} Ω from your summing point to the rest of the world . . . " Then, I demonstrate that the measured impedance is typically a lot more than the specifications say, but I agree that I wouldn't dare count on that fact. So, I guard the summing point to ground with a grounded foil surrounding the critical nodes on *both* the top and bottom of the boards.

With the addition of these grounds, the circuit can perform well even under worst-case humidity conditions. After all, the internal volume of the glass-epoxy insulator is always dry, whereas the surface is the place where you can easily get a leakage problem due to dirt or moisture. That's where you have to prevent the leakage. Of course, crosstalk and high-frequency capacitive-coupling problems are caused by adjacent foil locations and are cured by the same guarding and shielding just discussed to prevent foil leakages. To help you plan a good layout, think about what dv/dt and di/dt will do in a poor layout.

Location, Location, and . . . Location

The field of PC-board layout is a subject unto itself. But there are some things you can do or add to a layout that will make testing the circuit much easier. Thoughtful designers have a store of these tricks, but I bet very few people write them down. In my world, the unwritten rules are the ones that are broken, so we are trying to write them down. I recommend that all designers write down a list of their good ideas. Some layout tips from my list are:

• Make sure that the signals you need access to, for troubleshooting or analysis, are easy to find and probe. Make a small hole in the solder mask for accessibility.

• Include a silk-screen layer of labels in your layout artwork showing each component and its reference designation. It's also a good idea to label numbered test nodes and the correct polarity of diodes and electrolytic capacitors.

• Arrange the signal paths so that if you are desperate, you can easily break a link and open a loop, be it analog or digital.

• Many modern PC boards have multiple layers and sophisticated patterns of ground planes, power-supply busses, and signal flows. Troubleshooting such a board requires specialized techniques and skills and all sorts of "maps," so you don't get lost or confused. Make sure that all the board's nodes are accessible, not hidden or buried on an inside layer, or under a large component.

• When possible, leave adequate space around components, especially ones that are more likely to fail and need to be replaced. Such components might be part of circuits that lead off the board and into the realm of ESD transients and lightning bolts, and thus might occasionally fail.

• Locate delicate components away from the edge of boards, where they might be damaged by rough handling.

• Beware of using eyelets to connect different layers of foil on your PC board.

Years ago, plated-through holes were considered risky, so we used eyelets to connect the top and bottom foils. When these eyelets went through temperature cycling, the thermal stresses would cause the eyelets to *lift* the foil right off the boards. Even in the last year, I have seen advertisements that sing out the praises of eyelets on PC boards, which scares the heck out of me. If you have to use an eyelet, don't count on it to connect the top foil to the bottom. These days, plated-through holes are quite reliable, but I still like to use two plated-through holes in parallel, whenever I have room, or to put a component's lead through the hole. It just makes me feel more comfortable. (See Chapter 13 for more comments on eyelets.)

More important than these layout conveniences is that your layout should not interfere with and, if possible, should enhance the expected performance of the circuit. I try to make my layouts so the PC runs (and the metal runs on an IC chip) are as short and compact as possible, especially the sensitive ones that would receive a lot of noise or leakage or have a lot of capacitance, if they were longer. Otherwise, you have a lot of long wire runs that turn into a whole hank of spaghetti.

On the other hand, sometimes we *have to* locate some of the components in odd locations, for different reasons such as thermal or human interface—which leads to the "spaghetti syndrome." But sometimes you have to do it—it's a matter of engineering judgment, a matter of the trade-offs you have to make. Recently I was trying to design a 200-MHz counter, to start and stop in less than 5 ns (Ref. 1). I was able to lay out the fast 100,000-type ECL gates right close to each other, and with this close layout, I was able to gate the clock counter ON and OFF very quickly, faster than 3 ns. Most designers of digital circuits are aware that, with high-speed logic, you can't just run the fast signals "any old way." You have to treat these signal paths as transmission lines and route them carefully. So, whether with printed-circuit foils or wire-wrapped conductors or IC metal runs, most digital-circuit designers have learned to design the wiring of fast digital circuits so that they work well—to avoid ringing and crosstalk.

However, I have seen cases in which an experienced digital designer had to add a few linear circuits into one corner of a mostly digital PC board. If the designer makes bad guesses about how to wire an op amp, the linear circuits may oscillate or exhibit bad cross-talk or work poorly. And, the availability of wire-wrap connections makes it tempting for the board designer to make a neat-looking layout with all the op amps and comparators and feedback resistors and capacitors in neat rows. *Unfortunately*, this neatness causes some critical interconnections to be a couple of inches apart, and causes other signals to be bussed too closely to each other. And then the designer is puzzled as to why the amplifiers and comparators are oscillating so badly.

So, PC-board designers should be made aware that the layout of linear ICs can be quite critical. The 2-inch spacing that you would *never* allow, for example, between a digital IC and *its* bypass capacitor is the same 2-inch spacing that makes an op amp unhappy when its inverting or noninverting input has to travel that distance to various resistors or capacitors. As I will explain in future chapters on active-circuit troubleshooting, there are good reasons to keep those summing-point foils short, neat, and compact. Meanwhile it is not fair to assume that the PC board layout-designer will be able to guess which nodes are critical. The circuit-design engineer must provide a list of the nodes that are critical or sensitive—a list of large, noisy signals to keep away from delicate inputs, and so forth. It's only reasonable.

Some engineers like to use narrow PC foil runs, as narrow as possible. Others like to use wide foils and narrow stripes of insulation. Neither one is wrong, but you should be aware of the advantages of using large areas of foil when you are laying out a high-power transistor or IC. The collector of a TO-92 transistor can put a lot of heat down its (copper) collector lead, and an extra square inch of PC foil (on either or on both sides of the PC board) can spread that heat out and help keep the transistor cool. The same is true for high-power ICs: if you look at the data sheet of a medium-power IC such as the LM384, the curves show that 2 square inches of copper foil can help keep the IC much cooler than minimum foil, and 6 square inches is even better, when the device's 6 ground pins carry the heat out of the package into the foil. But some people point out that leaving too much foil on a PC board can cause warping after wave-soldering.

Engineers often assume that a printed-circuit trace has virtually no resistance and no IR drop. But, when you run large currents through a foil run, you will be unpleasantly surprised by the IR drop you'll see. The classic example is a layout where the signal ground for a preamplifier is mixed and shared with the ground-return path for the power supply's bridge and filter capacitor. This return path will see ampere-size surges 120 times per second. Needless to say, that preamplifier won't have "low noise" until the path for the current surges is essentially divorced from the preamp's ground. For precision work, your PC-board layout must include well-thought-out ground paths for your sensitive circuitry.

When you think about it, separating power-supply grounds that carry lots of current from voltage-sensing circuitry is similar to Kelvin connections, which are commonly used in test instruments. A Kelvin connection uses four wires: one pair of leads is meant to carry current and the other pair senses voltage across a device. Keeping the idea of Kelvin connections in mind when designing your PC board will help you optimize your grounding scheme. In fact, when I draw up my schematic, I label each set of grounds separately. If it has a lot of dirty currents flowing, I keep that separate as Power Ground; if it has to be especially clean, that's a High-Quality ground, indicated with a triangular ground symbol. Then the grounds will be connected together at only one point.

I don't have very much to say about printed-circuit boards for surface-mount devices, because I have not worked on them, and I am not an expert about them. I hear that they are, of course, more challenging, and require more meticulous work in every way. In other words, if you are an expert at engineering ordinary PC boards, you can probably study really hard and do it well. If you are not an expert, well, it's not a good place to start. After you get your board built, one of the worst problems is that of thermal cycling and stresses. Packages such as LCCs (Leadless Chip Carriers) have problems because of their zero lead length—they have no mechanical compliance. If the PC board material does not have the same thermal coefficient of expansion as the IC package, the soldered joints can be fatigued by the stresses of temp cycling, and may fail early. This is especially likely if you have to cycle it over a wide (military) temperature range—exactly where people wish they had perfect reliability. The commercial surface-mount ICs with gull-wing leads or J-leads are more flexible and more forgiving, and may cause less problems, but you have to do your homework. Too much solder and the leads get stiffened excessively; too little, and there's not enough to hold on.

Kelvin Connections Improve Measurement Accuracy

The usefulness of Kelvin contacts and connections is not widely appreciated. In fact when Julie Schofield, an Associate Editor at *EDN*, asked me some questions about them recently, I was surprised to find very little printed reference material on the subject. I looked in a few dozen reference books and text books and didn't find a decent definition or explanation anywhere. I did find some "Kelvin clips," which facilitate Kelvin-connected measurements, in a Keithley catalog. I also found some Textool socket data sheets, which mentioned, in a matter-of-fact way, the advantages of Kelvin contacts. I'll try to explain their usefulness in more detail here, since neither Julie nor I could find a good description in any technical encyclopedia.

Perhaps the most common use of Kelvin connections is the remote-sense technique. Kelvin connections and sockets let you bring a precise voltage right to the terminal or pin of your circuit under test. If you don't control the supply voltage precisely, you might start failing parts that are actually meeting their specs.

For example, let's say we want to test the load regulation of an LM323, a 5-V regulator, when V_{in} is held at exactly +8.00 V, and the load changes from 5 mA to 3.00 A (Fig 5.4). In this circuit, there are four pairs of Kelvin connections at work. The first pair is located at the power supply's output. This programmable supply's remote-sense terminals permit it to maintain an accurate 8.00-V output right up to the pin of the DUT. This is commonly called "remote sense," when you are in the power-supply business, but actually it represents a Kelvin connection. This is important because if the 8.00-V supply dropped to 7.9 or 7.8, that would be an unfair test.

The second Kelvin connection in Figure 5.4 is located at the output of the DUT. In order for you to observe the changes in V_{out} as you apply various loads, the Kelvin contacts provide Force leads for the 3 A of output current. They also provide Sense leads, so you can observe the DUT's output with a high-impedance voltmeter. Note that there are two Sense and two Force connections to the ground pin of the DUT. You don't really need all *four* contacts—you can tie both Force leads together and also both Sense leads together. You can do that because there is no significant current flowing in the Sense lead; and in the Force lead, we don't care how much current flows, nor do we care exactly what the voltage drop is.

The op amp in Figure 5.4 forces the DUT's output current through the Darlington transistor and then through a 0.1 Ω precision resistor. The only way to use a 0.1-Ω resistor with any reasonable accuracy and repeatability is to use the 4-wire (Kelvin) connections as shown. The op amp can force the upper Sense lead to be precisely 300 mV above the lower Sense level, even if the lower end of the resistor does rise above ground due to various IR drops in wires or connectors.

There are several places in this circuit that we could call ground, but the only ground we can connect to the 300-Ω resistor and get good results is the Sense lead at the bottom of the 0.1-Ω resistor. If you connect the bottom of the 300-Ω resistor to any other "ground," shifts in the IR drops would cause relatively large and unpredictable and unacceptable shifts in the value of the 3-A current—in other words, Inaccuracy and Trouble. So, when you're running large currents through circuits, think about the effects of IR drops in various connectors and cables. If you think IR drops will cause trouble, maybe Kelvin connections can get you out of it.

Figure 5.4's fourth Kelvin connection is hidden inside the LM323 5-V regulator, which has separate Force and Sense connections to the output terminal. A fifth Kelvin connection is also concealed inside the current-limit circuitry of the regulator. Here, the device senses the load current with a 4-wire, Kelvin-connected resistor and sends that voltage to the current-limit sense amplifier.

The use of Kelvin sockets is not confined to large power transistors or high-current circuits. Consider a voltage reference with 2 mA of quiescent current. If you're trying to observe a 1-ppm stable reference and the ground connection changes by 5 mΩ (which most socket manufacturers do not consider disastrous), the 10-μV shift that results from this change in ground impedance could confound your measurements. If you want to avoid trouble in precision measurements, avoid sockets or at least avoid sockets that do not have Kelvin contacts. Lord Kelvin—William Thomson before he was appointed a baron—did indeed leave us with a bag full of useful tools.

Avoid Cold-Soldered Joints

I have a few comments to add about solder; most of the time we take it for granted. You'll normally use ordinary rosin-core tin-lead solder. If you avoid jiggling the

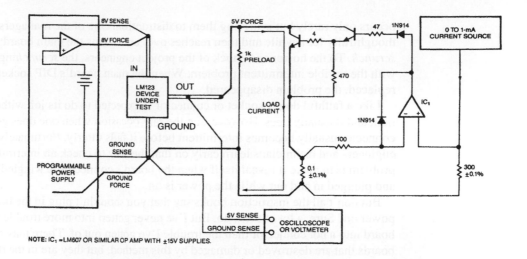

Figure 5.4. By using Kelvin connections you can avoid measurement errors caused by IR drops in the circuit that you're trying to measure, and in its connections. In this circuit, there are (at least) four pairs of Kelvin connections.

soldered joint as it is cooling, you won't get a cold-soldered joint. But you should know what a cold-soldered joint looks like and how much trouble one can cause in a critical circuit. I feel sort of sad that today's young people aren't building kits for electronic equipment. In the old days, you could learn all about cold-soldered joints before you got into industry, by building a "Heathkit" or a "Knightkit." I built several of each. I made a few cold-soldered joints and I had to fix them. With modern wave-soldering equipment, it's fairly easy to avoid cold-soldered joints in your production line. But on hand-soldered circuits, it's always a possibility to have cold-soldered joints, so, if you have a nasty problem, don't forget the old solution: Re-solder every joint. Once in a while you'll find a joint that never got any solder at all!

If for some reason you have acid-core solder around—it's mainly used for plumbing and is not found in most electronics labs, for good reason—keep it strictly labeled and segregated from ordinary rosin-core solder. Acid will badly corrode conductors. Also, keep specialty solders such as high-temperature solder, low-temperature solder, silver solder, and aluminum solder in a separate place, to avoid confusion. There is also solder for stainless steel, which requires special flux.

Recently I have heard people promoting silver-solder as a kind of superior solder for splicing speaker cables. The "Golden Ear" set claim that this solder makes the audio sound better. However, I must caution you that silver-solder requires rather high temperatures, such that you need a small torch, and some messy borax flux, and I suspect that the high temperatures will do a lot more damage to the insulation and to the copper wire (by oxidizing it excessively) than any advantage you might get from a "superior soldered joint."

Make Good Connections

Printed-circuit boards aren't the only assembled component you'll have to contend with while trying to make circuits work. In Tracy Kidder's Pulitzer-Prize-winning book, *The Soul of a New Machine* (Ref. 2), one of the crucial moments occurs when the engineers explain to a management team that their new computer has a flaw that occurs only rarely but is driving them to distraction. One of the managers stands thoughtfully for a while and then reaches over and warps the main board: *Scrunch, scrunch.* To the horror and shock of the project engineers, the *scrunch*ing correlates with the terrible intermittent problem. When the main board's DIP sockets were replaced, the problem disappeared.

Like a faithful dog, a socket or connector is expected to do its job without question, and it usually does. However, on the rare occasion when one does go bad, the connector usually becomes intermittent before it fails utterly. Fortunately, many engineers and technicians learn early on that the way to check an intermittent problem is to make it reveal itself when the board's connector is wiggled and jiggled and plugged in and out while the power is on.

But don't all the instruction books say that you shouldn't plug in the board with the power on? Sure, a lot of them do. But I've never gotten into more trouble plugging a board into a hot connector than the trouble I've gotten out of. There may be some boards that are destroyed or damaged by this method, but they are in the distinct minority and should be studied. One way to help avoid problems is to make the ground fingers on a printed-circuit edge connector stick out longer than the other fingers. Thus, ground will be established before any other connection. Still, if you have a board that tends to latch up because the power-supply sequencing may be improper, you have to be prepared to stop plugging the board into that hot socket—fast.

Learn by Fiddling and Tweaking

There are many situations that can foul things up, but we all tend to learn more from fiddling around with things, tweaking and unplugging, than by purely cerebral processes. Once I had a technician who thought that DIP sockets should not be secured in place by tack soldering, but by glue. This technique worked fine for a while, but occasionally the sockets would act like an open circuit on one pin or another. To solve the problem, we used an old technique: We traced the circuit coming into the IC, and it was fine. We traced the signal coming out of the IC; nothing. Then, we traced the signals on the pins of the DIP itself; the signals were not the same as the signals on the socket, not at all. I finally realized that the glue was getting into the internal voids of the socket and preventing the IC's pin from making a true connection.

We banned the glue from that task, and the problem went away, mostly. Still, both before and after that time, we have seen sockets that just failed to connect to an IC's pin. You merely have to probe to the pin of the IC itself, not just to the socket, to nail down this possibility. Sometimes, the pin goes into the socket and actually fails to connect; but, more often than not, the pin is simply bent under the package.

There is one other kind of problem you can have with a socket, as a friend of mine recounted. He was trying to troubleshoot a very basic op-amp circuit, but its waveforms did not make sense. After several minutes, he turned his circuit over and realized he had forgotten to plug an op amp into the socket. This example leads us to McKenna's Law (named after an old friend, Dan McKenna): "You can't see it if you don't look at it." We invoke this law when we discover that we forgot to plug in a line cord or connect something. A vital part of troubleshooting is realizing that we are all at the mercy of McKenna's Law when we get absent-minded.

Connectors and sockets usually do more good than harm. They permit you to check options and perform experiments that may seem absurd and preposterous, yet are Instructional and life-saving. Once a friend was in the throes of a knock-down-drag-out struggle to troubleshoot a fast A/D converter. He had tried many experiments, but a speed problem eluded him. He asked me if he should try a socket for a critical high-speed component. At first, I was aghast. But, after I thought about it and realized that the socket would add barely 1 pF of capacitance, I said, "Well, OK, it may not do much harm."

The addition of the socket led to the realization that the speed problem was critically correlated with that component, and the problem was soon solved. The socket that might have caused terrible strays actually caused almost no harm and, in fact, facilitated the real troubleshooting process. If nothing you do leads in an encouraging direction, and you have a half-baked notion to tell your technician to install a socket, that may be the best idea you have all day. The socket may do very little harm and could lead to many experiments, which might give you the vital clue that puts you on the track of the real culprit.

When is a Connector Not a Connector?

When it's a relay. A relay is an electrically-controlled connector, and though relays are not as popular as they used to be, there are many times and many ways to use a relay to get a job done exactly right. Conversely, you can use a relay to get things done wrong, and you don't want to do that. Let's discuss.

Some relays are made with gold-plated or other precious-metal contacts, for the highest reliability in low-level circuits. What is the definition of low-level? As with every relay application, you have to refer to the manufacturer's data sheets. Very important. Because if you try to use a low-level relay for a high-power use, the contacts can erode, wear, or maybe even weld together. Conversely, if you try to use a heavy-duty relay, with its special metallurgy and tungsten contacts, they can run dry and refuse to contact at all in a high-impedance circuit. Gold is a wonderful (if expensive) material, and it can prevent contacts from running "dry," but it won't be found in high-current contacts.

Reed relays are hermetically sealed, and have good proven reliability if properly applied. I evaluated some recently that had less than 5 fA of leakage, when I guarded them carefully.[3] I was impressed. Also they can be engineered to have low thermocouples and fast response; but they need a lot of ampere-turns as there is no iron in their magnetic circuit, so they are not normally considered low-power devices.

Oh, yes, I must add in one more comment on reed relays. If you install a well-built coil around a clean glass reed, you can get those low leakages. But if you BUY a reed relay, complete with coil, all packaged in a neat package, I bet I can tell you what the package is made out of: nylon. At room temperature, nylon is a fair insulator, but at 35 degrees, under conditions of high humidity, nylon is a *LOUSY* insulator. You can't even get $10^9 \, \Omega$ of leakage. So, if you want decent low leakage, you may well have to "roll your own" reed relays.

All mechanical relays have contact bounce. This may last 2 or 20 ms, and the duration can vary. When you want relays or switches to talk to digital logic circuits, antibounce circuits are *de rigeur*. Also, when there is even a moderate amount of voltage *or* current, the manufacturer will usually spell out the need for some series RC network to put across the contacts, to help minimize arcing or "burning." You *gotta* read the data sheet.

But, these days, not all relays are mechanical. First of all, there are mercury-wetted relays which are credited with giving no contact bounce. Most of these must be held upright or they do not work. There *are* mercury-wetted reed relays that do not bounce

3. Coto Type 1240-12-2104, Coto Corporation, 55 Dupont Drive, Providence, Rhode Island 02907. (401) 943-2686.

and can be used in any position. But even they will not work at temperatures colder than −38 °C, where mercury freezes.

Then there are the solid-state relays. Some of them can switch many amperes—but they often use SCRs, and thus have a loss of more than a volt. You can't use *that* for a little signal. Others have low-ohm MOSFETs, and can handle a few amperes with low losses and low offset voltage. But the big ones have a lot of leakage and capacitance (which is not always mentioned). The little ones are nice and delicate, for precision switching, but cannot carry many milliamperes.

So, for high reliability, you have to be pretty knowledgeable and thoughtful, and selective, when you choose a relay, or you'll pick one that's inappropriate for your application, and you'll be embarrassed when some of them provide poor performance, or fail sooner than expected.

When Is a Relay Not a Relay?

When it is just a switch, mechanically operated by hand. But when you choose a switch, the contacts have almost exactly the same limitations as the relay's contacts. There are high-current ones, there are delicate ones, there are hermetically-sealed ones. So in the same way, be thoughtful when you make your choice. You *do* have the advantage here that if you try to wear out a relay, a few million operations can cause failures in just a few weeks; but most people couldn't wear out a switch by hand fast enough to get in trouble! As with relays, if you aren't sure what the data sheet is trying to tell you, talk to the manufacturer's good people for advice and interpretation. They may have a switch in their "back room" that is just what you need.

Weird Wired World

Now, I'll add a few pithy comments about wire and cable. Not all wire is the same. For example, when I first got a job in electronics, I was having a lot of trouble with Teflon-insulated wire. The wires would often break right at the point where the solder stopped. After several engineers assured me that all wire was the same and suggested that I was just imagining things, I was ready to scream.

Finally, I found an engineer who admitted that cable manufacturers *couldn't* put individually tinned wires into a Teflon insulator, as they do with plastic-insulated wire. At the temperatures at which the Teflon is extruded, the solder would all flow together, thus making the stranded wire a solid wire. Instead, cable manufacturers use silver-plated wire strands for Teflon-insulated wire. With this type of wire, solder tends to wick up into the strands, thus making the wire quite brittle. Once I understood the wire's structure, I was able to solve my problems by adding strain relief for any bends or pulling stresses.

As I mentioned in Chapter 2, the ordinary plastic-insulated single-conductor wire that is used in telephones has just the right stiffness to make good twisted-pair wire for making capacitors with values of 1, 2.1, or 4.95 pF. The wire doesn't have a Teflon dielectric, but it's good enough for most applications.

Consider Your Wire Type

Shielded or coaxial cables, such as RG-58U, RG-174, shielded twisted pairs, and other special flat cables, all have their place in the job of getting signals from *here* to *there* without undue attenuation or crosstalk. When you have a large number of wires mindlessly bundled together and you don't have any bad crosstalk, you're witnessing a miracle. Often you have to unbundle the wires and separate the offending ones or the sensitive, delicate signals from the rest. Also, you may end up rewiring some or all of the wires into shielded cables.

Remember, Teflon is a good insulator, but air is even better. If you have to add struts, standoffs, or spacers to make sure that the critical wires stay put, go right ahead. If you have problems, the wire manufacturers can give you some advice.

Conversely, just as you use Teflon or air when you need a superior insulator, you have to be careful to get your best conductances. A friend who is an amateur radio operator says that many kinds of problems in RF circuits arise because nuts and bolts are used to make ground connections. If a lock washer or star washer is not included, the mechanical connection can loosen, the ground impedances will change with every little stress or strain, and nasty intermittent electrical problems will result. So, a major factor in the reliability of these circuits is ensuring the integrity of all bolted joints by always including star washers. And, make sure that wires and connectors do not get so loose as to hurt the reliability of your circuit or system. (See also comments on star washers in Chapter 13.)

When you use shielded cable, should you ground the shield at one end or at both ends? Many cases call for a ground at the receiving end of the cable, but there are cases in which the shield is the main ground return. Neither way is necessarily bad, but be consistent. Likewise, in the design and the execution of the design, avoid ground loops, which can cause weird noise problems. In my systems, I build my analog ground system completely separate from the digital ground and make sure that the case or package ground is also strictly divorced. Then, after I use an ohmmeter to confirm that these grounds are really separate, I add one link from the analog ground to the digital ground and another link to the case or chassis. This technique works well for me, and I recommend it.

It is a little-known fact that some coaxial cable can degrade just sitting on the shelf. (Well, that's true, but it degrades *faster* if the shelf is sitting in the sunshine, or out in the rain. . . .) Some specialty types of cable whose codes and specifications are nominally similar can have an outer jacket that is not guaranteed to have good chemical stability. The jacket may be especially resistant to some chemicals, but less resistant to others. Specifically, in the 1950s there was a lot of military-surplus cable *similar to* RG-58 and RG-74 that did not have good stability. As the outer jacket degraded, the inner insulator was chemically degraded and the cable's UHF attenuation was degraded. In other cases, the outer shield was chemically attacked and corroded, and its conductivity got worse and its UHF attenuation was also impaired. Most of that old cable has died and you can't even find it in junk-piles any more. But there are still specialty cables being made and sold now that do not last as well or age as gracefully as you would expect a good wire to do. If you select a cable to be especially resistant to one kind of chemical, it may be less resistant than normal to the attack of other ordinary chemicals. So, you should be aware that, even in something as simple as a wire, there may be more problems than meet the eye.

Heck, I just had a couple yards of Teflon-insulated wire sitting outside my kitchen window, running over to a sensor for an electronic thermometer. The wire would only get, on the average, 1 hour per day of direct sunlight. After 10 years, the yellow wire was still in good shape, but the white insulation had just about died utterly. Who wants to explain that one?

Recently an engineer showed me the results of a study of wire for loudspeakers. He showed that the inductance of ordinary two-conductor wire (per 20 feet) can cause a small but noticeable phase shift—perhaps 10 degrees at 20 kHz, even with the large and ultra-expensive speaker cables ($10 per foot and up). But when he took flat "ribbon" cable with 40 conductors (which is typically used to bus digital signals

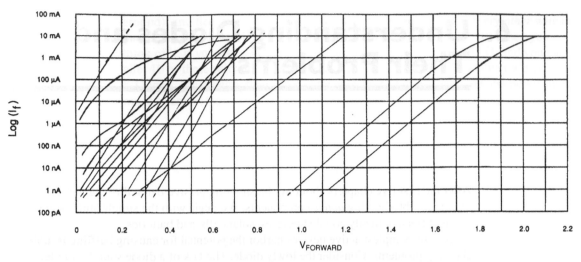

Figure 6.1. The diode made up of a transistor's emitter has high conductance over a wide range of currents. All the other diodes you can buy have inferior conductances, and they are just about all different.... surprise. For details, see Appendix E.

rectifiers are also available; they have been designed for fast switching-regulator and other high-frequency applications. They don't have quite as low V_Fs as Schottky diodes and are not quite as fast, but they are available with high reverse-voltage ratings and thus are useful for certain switch-mode circuit topologies that impress large flyback voltages on diodes.

When you reverse-bias these various diodes, ah, that is where you start to see even more wild dissimilarities. For example, the guaranteed reverse-current specification, I_{REV}, for many types of diodes is 25 nA max at 25 °C. When you measure them, many of these devices actually have merely 50 or 100 pA of leakage. But the popular 1N914 and its close cousin, the 1N4148, actually *do* have about 10 or 15 nA of leakage at room temperature because of their gold doping. So although these diodes are inexpensive and popular, it's wrong to use them in low-leakage circuits since they're much leakier than other diodes with the same leakage specs.

Why, then, do some low-leakage diodes have the same mediocre 25-nA leakage spec as the 1N914? Diode manufacturers set the test and price at the level most people want to pay, because automatic test equipment can test at the 25-nA level—but no lower—without slowing down. If you want a diode characterized and tested for 100 pA or better, you have to pay extra for the slow-speed testing. Of course, high-conductance diodes such as Schottkys, germaniums, and large rectifiers have much larger reverse leakage currents than do signal diodes, but that's not normally a problem.

If you want a very-low-leakage diode, use a transistor's collector-base junction instead of a discrete diode (Ref. 1). The popular 2N930 or 2N3707 have low leakage, typically. Some 2N3904s do, too, but some of these are gold-doped and are leakier. The plastic-packaged parts are at least as good as the TO-18 hermetic ones. You can easily find such "diodes" having less than 1 pA leakage even at 7 V, or 10 pA at 50 V. Although this low leakage is not guaranteed, it's usually quite consistent. However, this c-b diode generally doesn't turn ON or OFF very quickly.

Another source of ultra-low-leakage diodes are the 2N4117A and the PN4117A, -18 A, and -19 A. These devices are JFETS with very small junctions, so leakages well below 0.1 pA are standard with 1.0 pA max, guaranteed—not bad for a $0.40 part. The capacitances are small, too.

Speed Demons

When a diode is carrying current, how long does it take to turn the current off? There's another wide-range phenomenon. Slow diodes can take dozens and hundreds of microseconds to turn off. For example, the collector-base junction of a 2N930 can take 30 μs to recover from 10 mA to less than 1 mA, and even longer to the nanoampere level. This is largely due to the recombination time of the carriers stored in the collector region of the transistor. Other diodes, especially gold-doped ones, turn off *much* faster—down into the nanosecond region. Schottky diodes are even faster, much faster than 1 ns. However, one of my friends pointed out that when you have a Schottky diode that turns off pretty fast, it is still in parallel with a p-n junction that may still turn off slowly at a light current level. If a Schottky turns off from 4 mA in less than 1 ns, there may still be a few microamperes that do not turn off for a microsecond. So if you want to use a Schottky as a precision clamp that will turn off very quickly, as in a settling detector (Ref. 2), don't be surprised if there is a small long "tail."

Switching regulators all have a need for diodes and high-current rectifiers and transistors to turn off quickly. If the rep rate is high and the current large and the diode turns off slowly, it can fail due to overheating. You *don't* want to try a 1N4002 at 20 or 40 kHz, as it will work very badly, if at all. Sometimes if you need a moderate amount of current at high speed, you can use several 1N914s in parallel. I've done that in an emergency, and it seemed to work well, but I can't be sure I can recommend it as the right thing to do for long-term reliability. The right thing is to engineer the right amount of speed for your circuit. High-speed, fast-recovery, and ultrafast diodes are available. The Schottky rectifiers are even faster, but not available at high voltage breakdowns. When you start designing switching regulators at these speeds, you really must know what you are doing. Or at least, wear safety-goggles so you don't get hurt when the circuit blows up.

Turn 'Em Off—Turn 'Em On . . .

"Computer diodes" like the 1N914 are popular because they turn OFF quickly—in just a few nanoseconds—much faster than low-leakage diodes. What isn't well known is that these faster diodes not only turn OFF fast, they usually turn ON fast. For example, when you feed a current of 1.0 mA toward the anode of a 1N914 in parallel with a 40 pF capacitance (20 pF of stray capacitance plus a scope probe or something similar), the 1N914 usually turns ON in less than 1 ns. Thus, the V_F has only a few millivolts of overshoot.

But with some diodes—even 1N914s or 1N4148s from some manufacturers—the forward voltage may continue to ramp up past the expected DC level for 10 to 20 ns before the diode turns ON; this overshoot of 50 to 200 mV is quite surprising (Figure 6.2). Even more astonishing, the V_F overshoot may get worse at low repetition rates but can disappear at high repetition rates (Figure 6.2b–d).

I spent several hours once discovering this particular peculiarity when a frequency-to-voltage converter suddenly developed a puzzling nonlinearity. The trickiest part of the problem with the circuit's diodes was that diodes from an earlier batch had not exhibited any slow-turn-on behavior. Further, some diodes in a batch of 100 from one manufacturer were as bad as the diodes in Figures 6.2b and 6.2c. Other parts in that batch and other manufacturers' parts had substantially no overshoot.

When I confronted the manufacturers of these nasty diodes, they at first tried to deny any differences, but at length they admitted that they had changed some diffu-

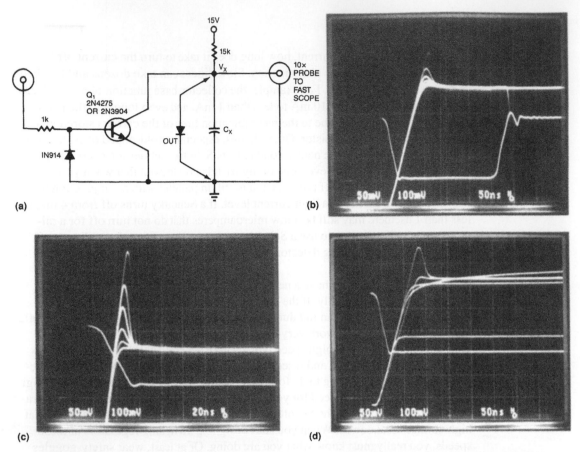

Figure 6.2. In this diode-evaluation circuit (a), transistor Q_1 simply resets V_x to ground periodically. When the transistor turns OFF, V_x rises to about 0.6 V at which point the diode starts conducting. In (b), when dV_x/dt is 8 V/μs, this 1N4148 overshoots as much as 140 mV at input frequencies below 10 kHz before it turns ON. At higher frequencies—120, 240, 480, 960, and 1920 kHz—as the repetition rate increases, the overshoot shrinks and disappears. Maximum overshoot occurs when f_{in} < 7 kHz. In (c), when dV_x/dt increases to 20 V/μs, this same 1N4148 overshoots as much as 450 mV at 7 kHz but only 90 mV at 480 kHz and negligible amounts at frequencies above 2 MHz. In (d), various diode types have different turn-on characteristics. The superimposed, 120-kHz waveforms are all invariant with frequency except for the bad 1N4148.

sions to "improve" the product. One man's "improvement" is another man's poison. Thus, you must always be alert for production changes that may cause problems. When manufacturers change the diffusions or the process or the masks, they may think that the changes are minor, but these changes could have a major effect on your circuit.

Many circuits, obviously, require a diode that can turn ON and catch, or clamp, a voltage moving much faster than 20 V/μs. Therefore, if you want any consistency in a circuit with fast pulse detectors (for example), you'll need to qualify and approve only manufacturers whose diodes turn ON consistently. So, as with any other unspecified characteristic, be sure to protect yourself against "bad" parts by first evaluating and testing and then specifying the performance you need. Also if you want to see

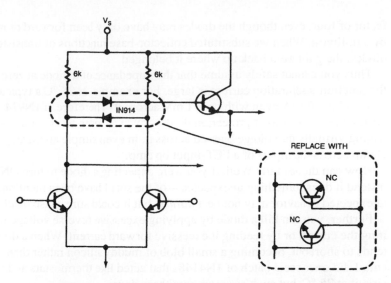

Figure 6.3. Even though the diodes in the first stage of this op amp are forward or reverse biassed by only a millivolt, the impedance of these diodes is much lower than the output impedance of the first stage or the input impedance of the second stage at high temperatures. Thus, the op amp's gain drops disastrously.

fast turn-on of a diode circuit, with low overshoot, you must keep the inductance of the layout small. It only takes a few inches of wire for the circuit's inductance to make even a good fast rectifier look bad, with bad overshoot.

One "diode" that does turn ON and OFF quickly is a diode-connected transistor. A typical 2N3904 emitter diode can turn ON or OFF in 0.1 nsec with negligible overshoot and less than 1 pA of leakage at 1 V, or less than 10 pA at 4 V. (This diode does, of course, have the base tied to the collector.) However, this diode can only withstand 5 or 6 V of reverse voltage, and most emitter-base junctions start to break down at 6 or 8 V. Still, if you can arrange your circuits for just a few volts, these diode-connected transistors make nice, fast, low-leakage diodes. Their capacitance is somewhat more than the 1N914's 1pF.

Other Strange Things That Diodes Can Do to You . . .

If you keep LEDs in the dark, they make an impressive, low-leakage diode because of the high band-gap voltage of their materials. Such LEDs can exhibit less than 0.1 pA of leakage when forward biassed by 100 mV or reverse biassed by 1 V.

Of course, you don't have to reverse-bias a diode a lot to get a leakage problem. One time I was designing a hybrid op amp, and I specified that the diodes be connected in the normal parallel-opposing connection across the input of the second stage to avoid severe overdrive (Figure 6.3). I thought nothing more of these diodes until we had the circuit running—the op amp's voltage gain was falling badly at 125 °C. Why? Because the diodes were 1N914s, and their leakage currents were increasing from 10 nA at room temperature to about 8 μA at the high temperature. And—remember that the conductance of a diode at zero voltage is approximately (20 to 30 mS/mA) $\times$ $I_{LEAKAGE}$. That means each of the two diodes really measured only 6 kΩ.

Because the impedance at each input was only 6 kΩ, the op amp's gain fell by a factor of four, even though the diodes may have only been forward or reverse biassed by a millivolt. When we substituted collector-base junctions of transistors for the diodes, the gain went back up where it belonged.

Thus you cannot safely assume that the impedance of a diode at zero bias is high if the junction's saturation current is large. For example, at 25 °C a typical 1N914 will leak 200 to 400 pA even with only 1 mV across it. Therefore, a 1N914 can prove unsuitable as a clamp or protection diode—even at room temperature—despite having virtually no voltage biassed across it, in even simple applications such as a clamp across the inputs of a FET-input op amp.

How can diodes fail? Well, if you were expecting a diode to turn ON and OFF, but instead it does something unexpected—of the sort I have been mentioning—that unexpected behavior may not be a *failure*, but it could sure cause *trouble*.

Further, you can kill a diode by applying excessive reverse voltage without limiting the current or by feeding it excessive forward current. When a diode fails, it tends to short out, becoming a small blob of muddy silicon rather than an open circuit. I did once see a batch of 1N4148s that acted like thermostats and went open-circuit at 75 °C, but such cases are rare these days.

One of the best ways to kill a diode is to ask it to charge up too big a capacitor during circuit turn-on. Most rectifiers have maximum ratings for how much current they can pass, on a repetitive and on a nonrecurring basis. I've always been favorably impressed by the big Motorola (Phoenix, AZ) books with all the curves of safe areas for forward current as a function of pulse time and repetition rate. These curves aren't easy to figure out at first, but after a while they're fairly handy tools.

Manufacturers can play tricks on you other than changing processes. If you expect a diode to have its arrow pointing toward the painted band (sometimes called the cathode by the snobbish) and the manufacturer put the painted band on the wrong end, your circuit won't work very well. Fortunately reverse-marked diodes are pretty rare these days. But just this morning, I heard an engineer call the "pointed" end of the diode an anode, which led to confusion and destruction. Sigh . . . .

Once I built a precision test box that worked right away and gave exactly the right readings until I picked up the box to look at some waveforms. Then the leakage test shifted *way* off zero. Every time I lifted up the box, the meter gave an indication; I thought I had designed an altimeter. After some study, I localized the problem to an FD300 diode, whose body is a clear glass DO-35 package covered with black paint. This particular diode's paint had been scratched a little bit, so when I picked up the test box, the light shone under the fixture and onto the diode. Most of these diodes didn't exhibit this behavior; the paint wasn't scratched on most of them.

To minimize problems such as the ones I have listed, I recommend the following strategies:

- Have each manufacturer's components specifically qualified for critical applications. This is usually a full-time job for a components engineer, with help and advice from the design engineer and consultation with manufacturing engineers.

- Establish a good relationship with each manufacturer.

- Require that manufacturers notify you when, or preferably before, they make changes in their products.

- Keep an alternate source qualified and running in production whenever possible.

My boss may gripe if I say this too loudly, but it is well known that having two good sources is better than having one. The argument that "One source is better than

two" falls hollow on my ears. Two may be better than seven or eight, but one is not
better than two.

Zener, Zener, Zener. . .

Just about all diodes will break down if you apply too much reverse voltage, but
zener diodes are *designed* to break down in a predictable and well-behaved way. The
most common way to have problems with a zener is to starve it. If you pass too little
current through a zener, it may get too noisy. Many zeners have a clean and crisp
knee at a small reverse-bias current, but this sharp knee is not guaranteed below the
rated knee current.

Some zeners won't perform well no matter how carefully you apply them. In contrast to high-voltage zeners, low-voltage (3.3 to 4.7 V) zeners are poor performers
and have poor noise and impedance specs and bad temperature coefficients—even if
you feed them a lot of current to get above the knee, which is very soft. This is because "zeners" at voltages above 6 V are really avalanche-mode devices and employ
a mechanism quite different from (and superior to) the low-voltage ones, which are
real zener diodes. At low-voltage levels, band-gap references such as LM336s and
LM385s are popular, because their performance is good compared with low-voltage
zeners.

Zener references with low temperature coefficients, such as the 1N825, are only
guaranteed to have low temperature coefficients when operated at their rated current,
such as 7.5 mA. If you adjust the bias current up or down, you can sometimes tweak
the temperature coefficient, but some zeners aren't happy if operated away from their
specified bias. Also, don't test your 1N825 to see what its "forward-conduction
voltage" is because in the "forward" direction, the device's temperature-compen-
sating diode may break down at 70 or 80 V. This break-down damages the device's
junction, degrades the device's performance and stability, and increases its noise.

The LM329 is popular as a 6.9-V reference because its TC is invariant of operating
current, as it can run from any current from 1 to 10 mA. The LM399 is even more
popular because of its built-in heater that holds the junction at +85°C. Consequently
it can hold 1/2 or 1 ppm per °C. The LM329 and LM399 types also have good long-
term stability, such as 5 or 10 ppm per 1000 hours, typically. The buried zeners in the
LM129/LM199/LM169 also have better stability than most discrete references
(1N825 or similar) when the references are turned on and off.

And before you subject a zener to a surge of current, check its derating curves
for current vs. time, which are similar to the rectifiers' curves mentioned earlier.
These curves will tell you that you can't bang an ampere into a 10-V, 1-W zener for
very long.

In fact, most rectifiers are rated to be operated strictly within their voltage ratings,
and if you insist on exceeding that reverse voltage rating and breaking them down,
their reliability will be degraded. To avoid unreliability, you can redesign the circuit
to avoid over-voltage, or you might add in an R-C-diode damper to soak up the energy; or you could shop for a controlled-avalanche rectifier. These rectifiers are rated
to survive (safely and reliably) repetitive excursions into breakdown when you exceed their rated breakdown voltage. The manufacturers of these devices can give you
a good explanation of how to keep out of trouble.

If you *do* need a zener to conduct a surge of current, check out the specially designed surge-rated zener devices—also called transient-voltage suppressors—from
General Semiconductor Industries Inc. (Tempe, AZ). You'll find that their 1-W devices, such as the IN5629 through IN5665A, can handle a surge of current better than

curve or list any realistic typical values; the sheets list only the worst-case values. Therefore, you may not realize that the V_F of an LED in an opto-isolator is a couple hundred millivolts smaller than that of discrete red or infrared LEDS. Conversely, the V_F of high-intensity, or high-efficiency, red LEDs tends to be 150 mV larger than that of ordinary red LEDs. (Refer to Appendix E.) And the V_F of DEADs (a DEAD is a **D**arkness **E**mitting **A**rsenide **D**iode; that is, a defunct LED) is not even defined.

Once I was troubleshooting some interruptor modules. In these modules, a gap separated an infrared LED and a phototransistor. An interruptor—say a gear tooth—in the gap can thus block the light. I tested one module with a piece of paper and nothing happened—the transistor stayed ON. What was that again? It turned out that the single sheet of paper could diffuse the infrared light but not completely attenuate it. A thin sheet of cardboard or two sheets of paper would indeed block the light.

Solar Cells

Extraneous, unwanted light impinging on the pn junction of a semiconductor is only one of many tricky problems you can encounter when you try to design and operate precision amplifiers—especially high-impedance amplifiers. Just like a diode's pn junction, a transistor's collector-base junction makes a good photodiode, but a transistor's plastic or epoxy or metal package normally does a very good job of blocking out the light.

When light falls onto the pn junction of any diode, the light's energy is converted to electricity and the diode forward biasses itself. If you connect a load across the diode's terminals, you can draw useful amounts of voltage and current from it. For example, you could stack a large number of large-area diodes in series and use them for recharging a battery. The most unreliable part of this system is the battery. Even if you never abuse them, batteries don't like to be discharged a large number of cycles, and your battery will eventually refuse to take a charge. These days one reads all sorts of marvelous hype about battery-powered cars, but the writers always ignore the terrible expense of replacing the batteries after just a few hundred cycles. They seem to be pretending that if they ignore that problem, it will go away . . . .

So much for the charms of solar-recharged batteries. It's much better to use a solar-powered night-light. Remember that one? A solar-powered night-light doesn't need a battery; it simply needs a 12,000-mile extension cord. To be serious, the most critical problem with solar cells is their packaging; most semiconductors don't have to sit out in the sun and the rain as solar cells do. And it's hard to make a reliable package when low cost is—as it is for solar cells—a major requirement.

In addition to packaging, another major trouble area with solar cells is their temperature coefficients. Just like every other diode, the V_F of a solar cell tends to decrease at 2 mV/°C of temperature rise. Therefore, as more and more sunlight shines on the solar cell, it puts out more and more current, but its voltage could eventually drop below the battery's voltage, whereupon charging stops. Using a reflector to get even more light onto the cell contributes to this temperature-coefficient problem. Cooling would help, but the attendant complications rapidly overpower the original advantage of solar cells' simplicity.

Assault and Battery

Lastly, I want to say a few things about batteries. The only thing that batteries have in common with diodes is that they are both two-terminal devices. Batteries are complicated electrochemical systems, and large books have been written about the charac-

Figure 6.6. With a solar-cell array, you can make electricity when the sun shines. (Photo copyright Peggi Willis.)

teristics of each type (Refs. 5–10). I couldn't possibly give batteries a full and fair treatment here, but I will outline the basics of troubleshooting them.

First, always refer to the manufacturer's data sheet for advice on which loads and what charging cycles will yield optimal battery life. When you recharge a nickel-cadmium battery, charge it with a constant current, not constant voltage. And be sure that the poor little thing doesn't heat up after it is nearly fully charged. Heat is the enemy of batteries as it is for semiconductors. If you're subjecting your battery to deep-discharge cycles, refer to the data sheet or the manufacturer's specifications and usage manual for advice. Some authorities recommend that you do an occasional deep discharge, all the way to zero; others say that when you do a deep discharge, some cells in the battery discharge before the others and then get reversed, which is not good for them. I cannot tell you who's correct.

Sometimes a NiCad cell will short out. If this happens during a state of low charge, the cell may stay shorted until you ZAP it with a brief burst of high current. I find that discharging a 470 μF capacitor charged to 12 V into a battery does a good job of opening up a shorted cell. If 470 μF doesn't do it, I keep a 3800 μF to do the job.

When you recharge a lead-acid battery, charge it to a float voltage of 2.33 V per cell. At elevated temperatures, you should decrease this float voltage by about 6 mV/°C; again, refer to the manufacturer's recommendations. When a lead-acid battery is deeply discharged (below 1.8 V per cell), it should be recharged right away or its longevity will suffer due to sulfation.

Be careful when you draw excessive current from a lead-acid battery; the good strong ones can overheat or explode. Also be careful when charging them; beware of the accumulation of hydrogen or other gases that are potentially dangerous or explosive.

And, please dispose of all dead batteries in an environmentally sound way. Call your local solid-waste-disposal agency for their advice on when and where to dispose of batteries. Perhaps some can be recycled.

Figure 6.7. Maintaining a healthy battery involves careful attention to charging, discharging, and temperature. (Photo copyright Peggi Willis.)

References

1. Pease, Robert A., "Bounding, clamping techniques improve on performance," *EDN*, November 10, 1983, p. 277.

2. Pease, Bob, and Ed Maddox, "The Subtleties of Settling Time," *The New Lightning Empiricist*, Teledyne Philbrick, Dedham MA, June 1971.

3. Pease, Robert A., "Feedback provides regulator isolation," *EDN*, November 24, 1983, p. 195.

4. Pease, Robert A., "Simple circuit detects loss of 4-20 mA signal," *Instruments & Control Systems*, March 1982, p. 85.

5. *Eveready Ni-Cad Battery Handbook*, Eveready, Battery Products Div., 39 Old Ridgebury Rd., Danbury, CT. (203) 794-2000.

6. *Battery Application Manual*, Gates Energy Products, Box 861, Gainesville, FL 32602. (1-800-627-1700). (Note: A sealed lead-acid and NiCd battery manual.)

7. Perez, Richard, *The Complete Battery Book*, Tab Books, Blue Ridge Summit, PA, 1985.

8. Small, Charles H., "Backup batteries," *EDN*, October 30, 1986, p.123.

9. Linden, David, Editor-in-Chief, *The Handbook of Batteries and Fuel Cells*, McGraw-Hill Book Co., New York, NY, 1984. (Note: The battery industry's bible.)

10. Independent Battery Manufacturers Association, *SLIG System Buyer's Guide*, 100 Larchwood Dr., Largo, FL 33540. (813) 586-1408. (Note: Don't be put off by the title; this book is the best reference for lead acid batteries.)

7. Identifying and Avoiding Transistor Problems

Although transistors—both bipolars and MOSFETs—are immune to many problems, you can still have transistor troubles. Robust design methods and proper assumptions regarding their performance characteristics will steer you past the shoals of transistor vexation and the rocks of transistor disasters.

Transistors are wonderful—they're so powerful and versatile. With a handful of transistors, you can build almost any kind of high-performance circuit: a fast op amp, a video buffer, or a unique logic circuit.

On the other hand, transistors are uniquely adept at causing trouble. For example, a simple amplifier probably won't survive if you short the input to the power supplies or the output to ground. Fortunately, most op amps include forgiving features, so that they can survive these conditions. When the μA741 and the LM101 op amps were designed, they included extra transistors to ensure that their inputs and outputs would survive such abuse. But an individual transistor is vulnerable to damage by excessive forward or reverse current at its input, and almost every transistor is capable of melting. So it's up to us, the engineers, to design transistor circuits so that the transistors do not blow up, and we must troubleshoot these circuits when and if they do.

A simple and sometimes not-so-obvious problem is installing a transistor incorrectly. Because transistors have three terminals, the possibility of a wrong connection is considerably greater than with a mere diode. Small-signal transistors are often installed so close to a printed-circuit board that you can't see if the leads are crossed or shorted to a transistor's can or to a PC trace. In fact, I recall some boards in which the leads were often crossed and about every tenth transistor was the wrong gender— pnp where an npn should have been, or vice versa. I've thought about it a lot, and I can't think of any circuits that work equally well whether you install a transistor of the opposite sex. So, mind your Ps and Qs, your Ps and Ns, your 2N1302s and 2N1303s, and your 2N3904s and 2N3906s.

In addition to installing a transistor correctly, you must design with it correctly. First of all, unless they are completely protected from the rest of the world, transistors require input protection. Most transistors can withstand dozens of milliamperes of forward base current but will die if you apply "only a few volts" of forward bias. One of my pet peeves has to do with adding protective components. MIL-HDBK-217 has always said that a circuit's reliability decreases when components are added. Yet when you add resistors or transistors to protect an amplifier's input or output, the circuit's reliability actually improves. It just goes to show that you can't believe everything you read in a military specification. For detailed criticism of the notion of computing reliability per MIL-HDBK-217, see Ref. 1.

Similarly, if you pump current *out of* the base of a transistor, the base-emitter junction will break down or "zener." This reverse current—even if it's as low as nanoamperes or very brief in duration—tends to degrade the low-current beta of the transistor, at least on a temporary basis. So in cases where accuracy is important, find a way to avoid reverse-biassing the inputs. Bob Widlar reminded me that the high-current beta of a transistor is generally not degraded by this zenering, so if you are hammering the V_{EB} of a transistor in a switch-mode regulator, that will not necessarily do it any harm, nor degrade its high-current beta.

Transistors are also susceptible to ESD—electrostatic discharge. If you walk across a rug on a dry day, charge yourself up to a few thousand volts, and then touch your finger to an npn's base, it will probably survive because a forward-biassed junction can survive a pulse of a few amperes for a small part of a microsecond. But, if you pull up the emitter of a grounded-base NPN stage, or the base of a PNP, you risk reverse-biassing the base-emitter junction. This reverse bias can cause significant damage to the base-emitter junction and might even destroy a small transistor.

When designing an IC, smart designers add clamp diodes, so that any pin can survive a minimum of + and –2000 V of ESD. Many IC pins can typically survive two or three times this amount. These ESD-survival design goals are based on the "human-body" model, in which the impedance equals about 100 pF in series with 1500 Ω. With discrete transistors, whose junctions are considerably larger than the small geometries found in ICs, ESD damage may not be as severe. But in some cases, ESD damage can still happen. Delicate RF transistors such as 2N918s, 2N4275s, and 2N2369s sometimes blow up "when you just look at 'em" because their junctions are so small.

Other transistor-related problems arise when engineers make design assumptions. Every beginner learns that the V_{BE} of a transistor decreases by about 2 mV per degree Celsius and increases by about 60 mV per decade of current. Don't forget about the side effects of these rules, or misapply them at extreme temperatures. Don't make sloppy assumptions about V_{BE}s. For instance, it's not fair to ask a pair of transistors to have well-matched V_{BE}s if they're located more than 0.1 in. apart and there are heat sources, power sources, cold drafts, or hot breezes in the neighborhood. Matched pairs of transistors should be glued together for better results. Of course, for best results, monolithic dual transistors like the LM394 give the *best* matching.

I've seen people get patents on circuits that don't even work—based on misconceptions of the relationships between V_{BE} and current. It's fair to assume that two matched transistors with the same V_{BE} at the same small current will have about the same temperature coefficient of V_{BE}. But you wouldn't want to make any rash assumptions if the two transistors came from different manufacturers or from the same manufacturer at different times. Similarly, transistors from different manufacturers will have different characteristics when going into and coming out of saturation, especially when you're driving the transistors at high speeds. In my experience, a components engineer is a very valuable person to have around and can save you a lot of grief by preventing unqualified components from confusing the performance of your circuits.

Another assumption engineers make has to do with a transistor's failure mode. In many cases, people say that a transistor, like a diode, fails as a short circuit or in a low-impedance mode. But unlike a diode, the transistor is normally connected to its leads with relatively small lead-bond wires; so if there's a lot of energy in the power supply, the short circuit will cause large currents to flow, vaporizing the lead bonds. As the lead bonds fail, the transistor will ultimately fail as an *open* circuit.

More Beta—More Better?

It's nice to design with high beta transistors, and, "if some is good, more's better." But, as with most things in life, too much can be disastrous. The h-parameter, h_{rb}, is equal to $\Delta V_{BE} / \Delta V_{CB}$ with the base grounded. Many engineers have learned that as beta rises, so does h_{rb}. As beta rises and h_{rb} rises, the transistor's output impedance decreases; its Early voltage falls; its voltage gain decreases; and its common-emitter breakdown voltage, BV_{CEO}, may also decrease. (The Early voltage of a transistor is the amount of V_{CE} that causes the collector current to increase to approximately two times its low-voltage value, assuming a constant base drive. V_{Early} is approximately equal to 26 mV $\times$ (1/h_{rb})). So, in many circuits there is a point where higher beta simply makes the gain lower, *not* higher.

Another way to effectively increase "beta" is to use the Darlington connection; but the voltage gain and noise may degrade, the response may get flaky, and the base current may decrease only slightly. When I was a kid engineer, I studied the ways that Tektronix made good use of the tubes and transistors in their mainframes and plug-ins. Those engineers didn't use many Darlingtons. To this day, I keep learning more and more reasons not to use Darlingtons or cascaded followers. For many years, it's been more important (in most circuits) to have matched betas than to have sky-high betas. You can match betas yourself, or you can buy monolithic dual matched transistors like the LM394, or you can buy four or five matched transistors on one monolithic substrate, such as an LM3045 or LM3086 monolithic transistor array.

One of the nice things about bipolar transistors is that their transconductance, g_m, is quite predictable. At room temperature, $g_m = 38.6 \times I_C$ (This is much more consistent than the forward conductance of diodes, as mentioned in the previous chapter.) Since the voltage gain is defined as $A_V = g_m \times Z_L$, computing it is often a trivial task. You may have to adjust this simple equation in certain cases. For instance, if you include an emitter-degeneration resistor, R_e, the effective transconductance falls to $1/(R_e + g_m^{-1})$. A_V is also influenced by temperature changes, bias shifts in the emitter current, hidden impedances in parallel with the load, and the finite output impedance of the transistor. Remember—higher beta devices can have *much* worse output impedance than normal.

Also be aware that although the transconductance of a well-biassed bipolar transistor is quite predictable, beta usually has a wide range and is not nearly as predictable. So you have to watch out for adverse shifts in performance if the beta gets too low or too high and causes shifts in your operating points and biasses.

Field Effect Transistors

For a given operating current, field-effect transistors normally have much poorer g_m than bipolar transistors do. You'll have to measure your devices to see how much lower. Additionally, the V_{GS} of FETs can cover a very wide range, thus making them harder to bias than bipolars.

JFETs (Junction Field-Effect Transistors) became popular 20 years ago because you could use them to make analog switches with resistances of 30 Ω and lower. JFETs also help make good op amps with lower input currents than bipolar devices, at least at moderate or cool temperatures. The BiFET™ process[1] made it feasible to make JFETs along with bipolars on a monolithic circuit. It's true that the characteristic of the best BiFET inputs are still slightly inferior to the best bipolar ones in terms of V_{OS} temperature coefficient, long-term stability, and voltage noise. But these BiFET characteristics keep improving because of improved processing and innovative circuit design. As a result, BiFETs are quite close to bipolar transistors in terms of voltage accuracy, and offer the advantage of low input currents, at room temperature.

---

1. A trademark of National Semiconductor Corporation.

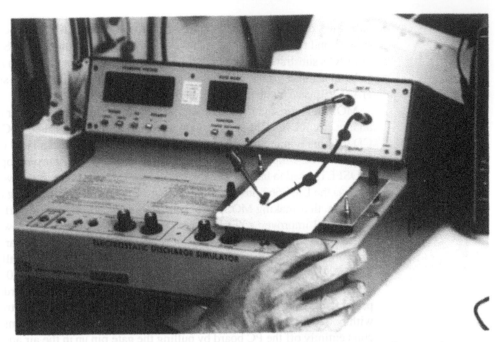

Figure 7.2. When you hit a component or circuit with a pulse of real ESD, you can never be sure what kind of trouble you'll get—unless you've already tested it with an ESD simulator. (Photo copyright Peggi Willis.)

cause enough I × R drop to force the entire emitter and its periphery to share the current. Now, let's halve the current and double the voltage: The amount of dissipation is the same, but the I × R drop is cut in half. Now continue to halve the current and double the voltage. Soon you'll reach a point where the ballasting (Figure 7.3) won't be sufficient, and a hot spot will develop at a high-power point along the emitter. The inherent decrease of V_{BE} will cause an increase of current in one small area. Unless this current is turned OFF promptly, it will continue to increase unchecked. This "current hogging" will cause local overheating, and may cause the area to melt or crater—this is what happens in "secondary breakdown." By definition you have exceeded the secondary breakdown of the device. The designers of linear ICs use ballasting, cellular layouts, and thermal-limiting techniques, all of which can prevent harm in these cases (Ref. 3). Some discrete transistors are beginning to include these features.

Fortunately, many manufacturers' data sheets include permitted safe-area curves at various voltages and for various effective pulse-widths. So, it's possible to design reliable power circuits with ordinary power transistors. The probability of an unreliable design or trouble increases as the power level increases, as the voltage increases, as the adequacy of the heat sink decreases, and as the safety margins shrink. For example, if the bolts on a heat sink aren't tightened enough, the thermal path degrades and the part can run excessively hot.

High temperature *per se* does not cause a power transistor to fail. But, if the drive circuitry was designed to turn a transistor ON and only a base-emitter resistor is available to turn it OFF, then at a very high temperature, the transistor will turn itself ON and there will be no adequate way to turn it OFF. Then it may go into secondary breakdown and overheat and fail. However, overheating does not by itself cause failure. I once applied a soldering iron to a 3-terminal voltage regulator—I hung it from the tip of the soldering iron—and then ran off to answer the phone. When I

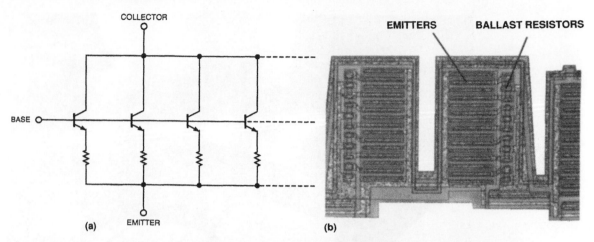

Figure 7.3. Ballast resistors, also known as sharing resistors, are often connected to the emitters of a number of paralleled transistors (a) to help the transistors share current and power. In an integrated circuit (b), the ballast resistors are often integrated with adjacent emitters. (Photo of National Semiconductor Corp's LM138.)

came back the next day, I discovered that the TO-3 package was still quite hot—+300 °C, which is normally recommended for only 10 seconds. When I cooled it off, the regulator ran fine and met spec. So, the old dictum that high temperature will necessarily degrade reliability is not always true. Still, it's a good practice to not get your power transistors that hot, and to have a base drive that can pull the base OFF if they do get hot.

You can also run into problems if you tighten the screws on the heat sink too tight, or if the heat sink under the device is warped, or if it has bumps or burrs or foreign matter on it. If you tighten the bolt too much, you'll overstress and warp the tab and die attach. Overstress may cause the die to pop right off the tab. The insulating washer under the power transistor can crack due to overstress or may fail after days or weeks or months. Even if you don't have an insulating washer, overtorqueing the bolts of plastic-packaged power transistors is one of the few ways a user can mistreat and kill these devices. Why does the number 10 inch-pounds max, 5 typ, stick in my head? Because that's the spec the Thermalloy man gave me for the 6/32 mounting bolts of TO-220 packages. For any other package, make sure you have the right spec for the torque. Don't hire a gorilla to tighten the bolts.

Apply the 5-Second Rule

Your finger is a pretty good heat detector—just be careful not to burn it with high voltages or very hot devices. A good rule of thumb is the 5-second rule: If you can hold your finger on a hot device for 5 seconds, the heat sink is about right, and the case temperature is about 85 °C. If a component is hotter than that, too hot to touch, then dot your finger with saliva and apply it to the hot object for just a fraction of a second. If the moisture dries up quickly, the case is probably around 100 °C; if it sizzles instantaneously, the case may be as hot as 140 °C. Alternatively, you can buy an infrared imaging detector for a price of several thousand dollars, and you won't burn your fingers. You will get beautiful color images on the TV screen, and contour maps of isothermal areas. You will learn a lot from those pictures. About twice a year, I wish I could borrow or rent one.

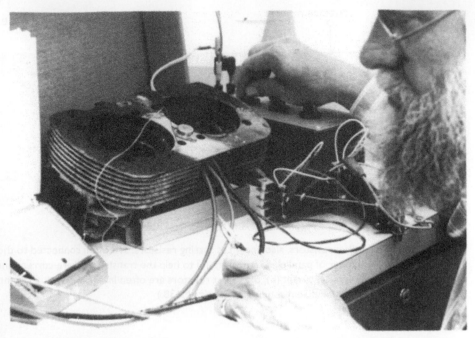

Figure 7.4. When using high-power amplifiers, there are certain problems you just never have if you use a big-enough heat sink. This heat sink's thermal resistance is lower than 0.5 °C/W. (Photo copyright Peggi Willis.)

Fabrication Structures Make a Difference

Another thing you should know when using bipolar power transistors is that there are two major fabrication structures: the epitaxial base, and the planar structure pioneered by Fairchild Semiconductor (Figure 7.5) (Ref. 4). (See my comments a couple paragraphs down concerning the obsolete single-diffused transistors.) Transistors fabricated with the epi-base structure are usually more rugged and have a wider safe-operating area. Planar devices feature faster switching speeds and higher frequency response, but aren't as rugged as the epi-base types. You can compare the two types by looking at the data sheets for the Motorola 2N3771 and the Harris 2N5039. The 2N5039 planar device has a current-gain bandwidth 10 times greater than the 2N3771 epi-base device. The 2N5039 also has a switching speed faster than the 2N3771 when used as a saturated switch, but the 2N3771 has a considerably larger safe area if used for switching inductive loads. You can select the characteristics you prefer, and order the type you need . . .

But be careful. If you breadboard with one type and then start building in production with the other, you might suddenly find that the bandwidth of the transistor has changed by a factor of 10 (or a factor of 0.1) or that the safe area doesn't match that of the prototypes. Also be aware that the planar power devices, like the familiar 2N2222 and 2N3904, are quite capable of oscillating at high frequencies in the dozens of megahertz when operated in the linear region, so you should plan to use beads in the base and/or the emitter, to quash the oscillation. The slower epi-base devices don't need that help very often.

When I first wrote these articles on troubleshooting back in 1988, you could still buy the older "single-diffused" transistors such as 2N3055H and the old 2N3771. I

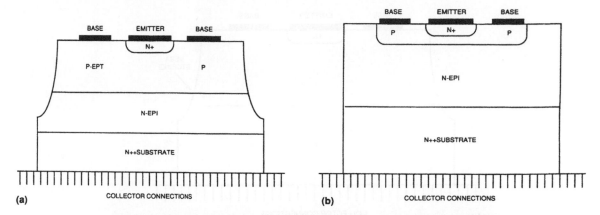

Figure 7.5. The characteristics of power transistors depend on their fabrication structure. The epitaxial-base structure (a) takes advantage of the properties of several different epitaxial layers to achieve good beta, good speed, low saturation, small die size, and low cost. This structure involves mesa etching, which accounts for the slopes at the die edges. Planar power transistors (b) can achieve very small geometries, small base-widths, and high-frequency responses, but they're less rugged than epitaxial-base types, in terms of Forward-Biassed SOA.

wrote all about how these devices had even more Safe Operating Area than the epi-base device, so you might want to order these if you wanted a "really gutsy" transistor for driving inductive loads. Unfortunately, these transistors were obsolescent and obsolete; they were slow (perhaps 0.5 MHz of fα), had a large die area, and were expensive. For example, although these transistors required only one diffusion, in some cases this diffusion had to run 20 hours. Because of all these technical reasons, sales shrank until, in the last 2 years, all the single-diffused power transistors have been discontinued.

So it's kind of academic to talk about the old single-diffused parts, (see Figure 7.6) but I included a mention here just for historical interest. Also, I included it because if you looked in my old *EDN* write-up and then tried to buy the devices I recommended, you would meet with incredulity. You might begin to question the sanity of yourself, or the salesman, or of Pease. When I inquired into the availability of these parts, I talked to many sales people who had *no idea* what I was talking about. Finally when I was able to talk to technical people, they explained why these transistors were not available—they admitted that I was not dreaming, but that the parts had been discontinued recently. These engineers at some of the major power-transistor manufacturers were quite helpful as they explained that newer geometries helped planar power transistors approach the safe area of the other older types without sacrificing the planar advantages of speed. Also, power MOSFETs had even wider amounts of SOA, and their prices have been dropping, and they were able to take over many new tasks where the planars did not have enough SOA. So the puzzle all fits together.

There is still one tricky problem. Originally the old 2N3771 was a single-diffused part. If you wanted to buy an epi-base part, that was the MJ3771. But now if you order a 2N3771, you get the epi-base part, which *does* meet and exceed the JEDEC 2N3771 specs. It just exceeds them a lot more than you would expect—like, the current-gain-bandwidth is 10 or 20 × higher. So, if you try to replace an *old* 2N3771 with a *new* 2N3771, please be aware that they are probably not very similar at all.

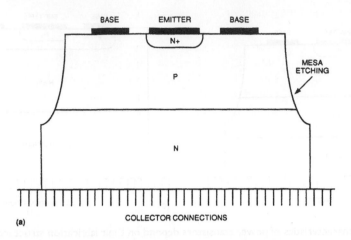

Figure 7.6. In the old single-diffused structure, n-type dopants were diffused simultaneously into the front and back of a thin p-type wafer. This structure produced rugged transistors with wider Safe Operating Areas than the more modern epitaxial-base transistor types, in terms of Forward-Biassed SOA. However, this fabrication has been *obsoleted*.

Power-Circuit Design Requires Expertise

For many power circuits, your transistor choice may not be as clear-cut as in the previous examples. So, be careful. Design in this area is not for the hotshot just out of school—there are many tricky problems that can challenge even the most experienced designers. For example, if you try to add small ballasting resistors to ensure current sharing between several transistors, you may still have to do some transistor matching. This matching isn't easy. You'll need to consider your operating conditions; decide what parameters, such as beta and V_{BE}, you'll match; and figure out how to avoid the mix-and-match of different manufacturers' devices. Such design questions are not trivial. When the performance or reliability of a power circuit is poor, it's probably not the fault of a bad transistor. Instead, it's quite possibly the fault of a bad or marginal driver circuit or an inadequate heat sink. Perhaps a device with different characteristics was inadvertently substituted in place of the intended device. Or perhaps you chose the wrong transistor for the application.

A possible scenario goes something like this. You build 10 prototypes, and they seem to work okay. You build 100 more, and half of them don't work. You ask me for advice. I ask, "Did they ever work right?" And you reply, "Yes." But wait a minute. There were 10 prototypes that worked, but the circuit design may have been a marginal one. Maybe the prototypes didn't really work all that well. If they're still around, it would be useful to go back and see if they had any margin to spare. If the 10 prototypes had a gain of 22,000, but the current crop of circuits has gains of 18,000 and fails the minimum spec of 20,000, your new units should not be called "failures." It's not that the circuit isn't working at all, it's just that your expectations were unrealistic.

After all, every engineer has seen circuits that had no right to work, but they did work—for a while. And then when they began to fail, it was obviously just a hopeless case. So, which will burn you quickest, a marginal design or marginal components? That's impossible to say. If you build in some safety margin, you may survive some of each. But you can't design with big margins to cover every possibility, or your design will become a monster. That's where experience and judgment must be invoked....

An old friend wrote to me from Japan, "Why do you talk about having to troubleshoot 40% of the units in a batch of switching regulators? In Japan that would be considered a bad design...." I replied that I agreed that it sounds like a problem, but until you see what is the cause of the problems, it is unfair to throw any blame around. What if it was a bad workmanship problem? Then that does not sound like a bad design—unless the design was so difficult to execute that the assembly instructions could not be followed. Or maybe a bad part was put in the circuit. Or maybe it was a marginally bad design and part of the circuit does need to be changed—perhaps an extra test or screening of some components—before the circuit can run in production. But you cannot just say that if there is ever trouble, it is the design engineer's fault. What if the design engineer designed a switching regulator that never had any problems in production—never ever—*but* it only puts out 1 W per 8 cubic inches, and all the parts are very expensive, and then there is a lot of expensive testing on each component before assembly, to prove that there is a good safety margin. Is that a good design? I doubt it. Because if you tried to build a plane with too big a safety factor, it might be bigger than a 747, but able to carry only 10 passengers. Every circuit should be built with an *appropriate* safety factor. If you only use a transistor that is always SURE to work well, that may be an uneconomic safety factor. Judgment is required to get the right safety factor.

MOSFETS Avoid Secondary Breakdown

When it comes to power transistors, MOSFETs have certain advantages. For many years, MOSFETs have been available that switch faster than bipolar transistors, with smaller drive requirements. And MOSFETs are inherently stable against secondary breakdown and current hogging because the temperature coefficient of I_{DS} vs. V_{GS} is inherently stable at high current densities. If one area of the power device gets too hot, it tends to carry less current and thus has an inherent mechanism to avoid running away. This self-ballasting characteristic is a major reason for the popularity of MOSFETs over bipolar transistors. However, recent criticism points out that when you run a MOSFET at high-enough voltages and low current, the current density gets very small, the temperature coefficient of I_{DS} vs. V_{GS} reverses, and the device's inherent freedom from current hogging may be lost (Ref. 5). So at high voltages and low current densities, watch out for this possibility. When the V_{DS} gets high enough, MOSFETS can exhibit current hogging and "secondary breakdown" similar to that of bipolars!!

The newer power MOSFETs are considerably more reliable and less expensive than the older devices. Even though you may need a lot of transient milliamps to turn the gate ON or OFF quickly, you don't need a lot of amps to hold it ON like you do with a bipolar transistor. You can turn the newer devices OFF quicker, too, if you have enough transient gate drive current available.

However, MOSFETs are not without their problem areas. If you persist in dissipating too many watts into a MOSFET, you can melt it just as you can melt a bipolar device. If you don't overheat a MOSFET, the easiest way to cause a problem is to forget to insert a few dozen or hundred ohms of resistance (or a ferrite bead) right at the gate lead of the device. Otherwise, these devices have such high bandwidths that they can oscillate at much higher frequencies than bipolar transistors.

For example, the first high-fidelity, all-MOSFET audio amplifier I ever saw blew up. It worked okay in the lab, but some misguided engineer decided that if a band-

whereas more modern amplifiers like the NSC OP-07 and the LM607 (gain = 6,000,000 min) have much less nonlinearity than older amplifiers.

Similarly, an op amp may have an offset-voltage temperature-coefficient specification of 1 μV/°C, but the op amp's drift may actually be 0.33 μV/°C at some temperatures and 1.2 μV/°C at others. Twenty or thirty years ago, battles and wars were fought over this kind of specsmanship, but these days, most engineers agree that you don't need to sweat the small stuff. Most applications don't require an offset drift less than 0.98 μV for each and every degree—most cases are quite happy when a 1 μV/°C op amp drifts less than 49 μV over 50 degrees.

Also, you don't often need to worry about bias current and its temperature coefficient, or the gain error's TC. If the errors are well behaved and fit inside a small box, well, that's a pretty good part.

There is one classical *caveat*, and I'll include it here because it never got mentioned in the EDN series. I showed my typed draft to 45 people, and then thousands of people looked at my articles, and *nobody* told me that I had forgotten this. *Namely*: If you run an op amp in a high-impedance circuit, so that the bias current causes significant errors when it flows through the input and feedback resistors, do *not* use the V_{OS} pot to get the circuit's output to zero. Example, if you have an LM741 as a unity-gain follower, with a source impedance of 500 kΩ and a feedback resistor of 470 kΩ, the 741's offset current of 200 nA (worst-case) could cause an output offset of 100 mV. If you try to use the V_{OS} trim pot to trim out that error, it won't be able to do it. If you have only 20 or 40 mV of this I × R error, you may be able to trim it out, but the TC and stability will be lousy. So, you should be aware that in any case where the I_{OS} × R is more than a few millivolts, you have a potential for bad DC error, and there's hardly any way to trim out those errors without causing other errors. When you get a case like this, unless you are willing to accept a crude error, then these errors are trying to tell you that you ought to be using a better op amp with lower bias currents.

And where did I find this reminder, not to trim out an I × R with a V_{OS} pot? There's a chapter in Analog Devices' *Data Converter Handbook*—it's in there (Ref. 1). Now, I've known about this trim problem for 25 years, but this isn't a problem that a customer asks us about even every year, these days, and I guess that's true for my colleagues, too. As it was not fresh in anybody's mind, well—we forgot to include it—we didn't notice it when it was missing. It just shows why you have to write things down!

An Uncommon Mode

A good example of misconstrued specs is the common-mode error. We often speak of an op amp as having a CMRR (Common Mode Rejection Ratio) of 100 dB. Does this number mean that the common-mode error is exactly one part in 100,000 and has a nice linear error of 10 μV per volt? Well, this performance is possible, but not likely. It's more likely that the offset-voltage error as a function of common-mode voltage is nonlinear (Figure 8.2). In some regions, the slope of ΔV_{OS} will be much better than 1 part in 100,000. In other regions, it may be worse.

It really bugs me when people say, "The op amp has a common-mode gain, A_{VC}, and a differential gain, A_{VD}, and the CMRR is the ratio of the two." This statement is silly business: It's *not* reasonable to say that the op amp has a differential gain or common-mode gain that can be represented by a single number. Neither of these gain numbers could ever be observed or measured with any precision or repeatability on any modern op amp. Avoid the absurdity of trying to measure a "common-mode gain of zero" to compute that your CMRR slope is infinite. You'll get more meaningful

Figure 8.1. If you run an op amp at such high impedances that $I_B \times R$ is more than 20 mV, you'll be generating big errors, and a V_{offset} trim-pot can't help you cancel them out. Please don't even try!

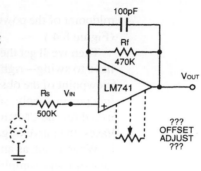

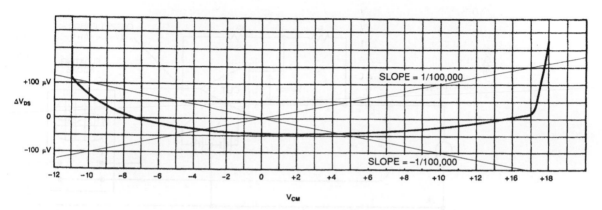

Figure 8.2. The CMRR of an op amp can't be represented by a single number. It makes more sense to look at the CMRR curve, ΔV_{OS} versus ΔV_{CM}, and note its nonlinearities, compared to a straight line with a constant slope of 1 part in 100,000.

results if you just measure the change in offset voltage, V_{OS}, as a function of common-mode voltage, V_{CM}, and observe the linear and nonlinear parts of the curve. What's a good way to measure the change in V_{OS} versus V_{CM}—the CMRR of an op amp? I know of a really good test circuit that works very well.

First, How *Not* to Test for CMRR

The first thing I always tell people is how *not* to measure CMRR. In Figure 8.3, if you drive a sine wave or triangle wave into point A, it seems like the output error, as seen by a floating scope, will be (N+1) times [V_{CM} divided by the CMRR]. But that's not quite true: you will see (N+1) times [the CM Error *plus* the Gain Error]. So, at moderate frequencies where the gain is rolling off and the CMRR is still high, you will see mostly the gain error, and your curve of CMRR vs. frequency will look just as bad as the Bode plot. That is because if you used the circuit of Figure 8.3, that's just what you will be seeing! There are still a few op-amp data sheets where the CMRR curve is stated to be the same as the Bode plot. The National LF400 and LF401 are two examples; next year we will correct those curves to show that the CMRR is actually much higher than the gain at 100 or 1000 Hz. National is *not*, by the way, the only company to have this kind of absurd error in some of their data sheets. . . .

Ah, let's avoid that floating scope—let's drive the sine wave generator into the midpoint of the power supply, and ground the scope and ground point A. (Figure 8.4.)

Then we'll get the true CMRR, because the output will stay near ground—it won't have to swing—right? Wrong! The circuit function has not changed at all; only the viewpoint of the observer has changed. The output *does* have to swing, referred to any power supply, so this still gives the *same wrong answer*. You may *say* that you asked for the CMRR as a function of frequency, but the answer you get is, in most cases, the curve of gain vs. frequency.

What about, as an alternative, the well-known scheme shown in Figure 8.5, where an extra servo amplifier closes the loop and does not require the op-amp output to do any swinging?

That's OK at DC—it is fine for DC testing, and for ATE (Automatic Test

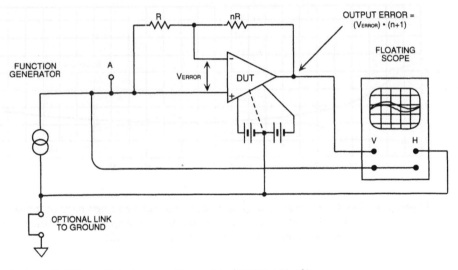

Figure 8.3. Is this a CMRR test? No, because $V_{error} = V_{cm}/CMRR + V_{out}/A_V$.

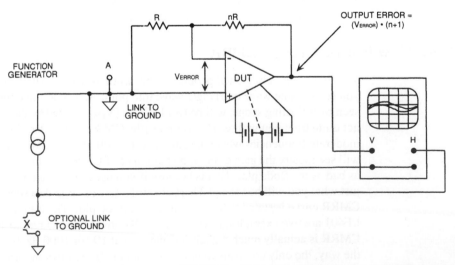

Figure 8.4. Is this any better than the previous "CMRR Test"? No, it's exactly the same! Still $V_{error} = V_{cm}/CMRR + V_{out}/A_V$.

Equipment), for production test, and for stepped DC levels. And it will give the same answer as my circuit at all low frequencies up to where it doesn't give the same answer. Now, what frequency would that be?? Nobody knows! Because if you have an op amp with low CMRR, the servo scheme will work accurately up to one frequency, and if you have an op amp with high CMRR, the servo scheme will work accurately only up to a different frequency. Also, the servo amplifier adds so much gain into the loop that ringing or overshoot or marginal stability at some mid frequencies is inevitable. That is much too horrible for me to worry about—I will just avoid that, by using a circuit which gives very consistent and predictable results.

Specifically, I ran an LF356 in the circuit of Fig. 8.3, and I got an error of 4 mV p-p at 1 kHz—a big fat quadrature error, 90 degrees out of phase with the output—see the upper trace in Figure 8.6.

If you think that is the CM error, you might say the CMRR is as low as 5,000 at 1 kHz, and falling rapidly as the frequency increases. But the actual CMRR error is about 0.2 mV p-p—see the lower trace of Figure. 8.6—and thus the CMRR is about

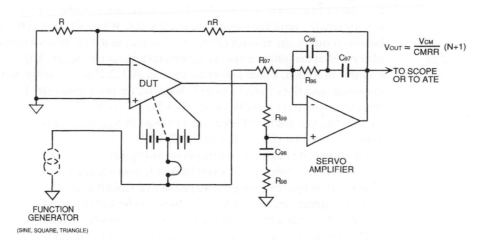

$$V_{OUT} = \frac{V_{CM}}{CMRR} (N+1)$$

Figure 8.5. This circuit is considered acceptable when used for DC CMRR tests in ATE systems. However, nobody ever tells you what are good values for the Rs and Cs, nor whether it is valid up to any particular frequency.

Figure 8.6. Trace (A) shows the "CMRR" error taken using the circuit of Figure 8.3. But it's not the CMRR error, it's really the gain error you are looking at, 4 mV p-p at 1 kHz. Trace (B) shows the actual common-mode error—about 1/20 the size of the gain error—measured using the circuit of Figure 8.7.

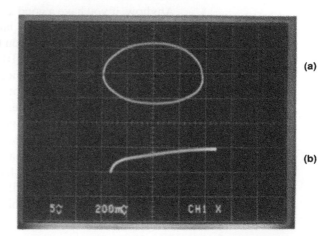

100,000 at 1 kHz or any lower frequency. Note also that, on this unit, the CM error is not really linear—as you get near –9 V, the error gets more nonlinear. (This is a –9-V/+12-V CM range on a 12-V supply; I chose a ±12-V supply so my function generator could over-drive the inputs.) So, the business of CMRR is not trivial— at least, not to do it right.

How to Do It Right . . .

As we discussed in the previous section, there are circuits that people use to try to test for CMRR, that do not give valid results. Just how, then, *can* we test for CMRR and get the right results??

Figure 8.7 is a darned fine circuit, even if I did invent it myself about 22 years ago. It has limitations, but it's the best circuit I've seen. Let's choose $R_1 = R_{11} = 1$ k, $R_2 = R_{12} = 10$ k, and $R_3 = 200$ k and $R_4 = a$ 500 Ω pot, single-turn carbon or similar. In this case, the noise gain is defined as $1 + [R_f/R_{in}]$, or about 11. See pages 100–101 for discussion of noise gain. Let's put a ±11-V sine wave into the signal input so the CM voltage is about ±10 V. The output error signal will be about 11 times the error voltage plus some function of the mismatch of all those resistors. Okay, first connect the output to a scope in cross-plot (X-Y) mode and trim that pot until the output error is very small—until the slope is nominally flat. We don't know if the CMRR error is balanced out by the resistor error, or *what*; but, as it turns out, we don't care. Just observe that the output error, as viewed on a cross-plot scope, is quite small. Now connect in R100a, a nice low value such as 200 Ω. If you sit down and compute it, the noise gain rises from 11 to 111. Namely, the noise gain *was* $(1 + R_2/R_1)$, and it then increases to $(1 + R_2/R_1)$ *plus* $(R_2 + R_{12})/R_{100}$. In this example, that is an increase of 100. So, you are now looking at a *change* of V_{out} equal to 100 times the input error voltage, (and that is V_{CM} divided by CMRR).

Of course, it is unlikely for this error voltage to be a linear function of V_{CM}, and that is why I recommend that you look at it with a scope in cross-plot (X-Y) mode. Too many people make a pretend game, that CMRR is constant at all levels, that CM error is a linear function of V_{CM}, so they just look at two points and assume every other voltage has a linear error; and that's just *too* silly. Even if you want to use some ATE (Automatic Test Equipment) you will want to look at this error at least three places—maybe at four or five voltages. Another good reason to use a scope in the X-Y mode is so you can use your eyeball to subtract out the noise. You certainly can't use an AC voltmeter to detect the CMRR error. For example, in Figure 8.6, the CM error is fairly stated as 0.2 mV p-p, not 0.3 mV p-p (as it might be if you used a meter that counted the noise).

Anyhow, if you have a good amplifier with a CMRR of about 100 dB, the CM error will be about 200 μV p-p, and as this is magnified by 100, you can easily see an output error of 20 mV p-p. If you have a really good unit with CMRR of 120 or 140 dB, you'll want to clip in the R100b, such as 20 Ω, and then the Δ (noise gain) will be 1000. The noise will be magnified by 1000, but so will the error and you can see what you need to see. *Now*, I shall not get embroiled in the question, are you trying to see exactly how good the CMRR really is, or just if the CMRR is better than the data-sheet value; in either case, this is the best way I have seen.

For use with ATE, you do not have to look with a scope, you can use a step or trapezoidal wave and look just at the DC levels at the ends or the middle or wherever you need. Note that you do not have to trim that resistor network all the time, nor do you have to trim it perfectly. All you have to know is that when the noise gain changes from a low value to a high value, and the output error changes, it is the

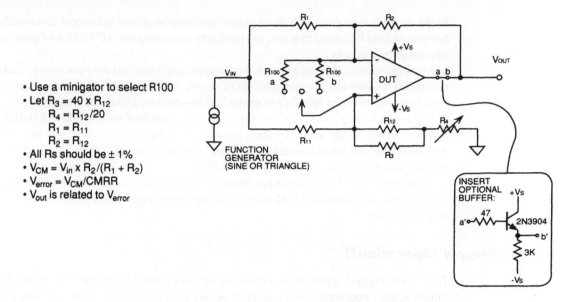

- Use a minigator to select R100
- Let $R_3 = 40 \times R_{12}$
    $R_4 = R_{12}/20$
    $R_1 = R_{11}$
    $R_2 = R_{12}$
- All Rs should be ± 1%
- $V_{CM} = V_{in} \times R_2/(R_1 + R_2)$
- $V_{error} = V_{CM}/CMRR$
- V_{out} is related to V_{error}

Figure 8.7. Here's how to evaluate CMRR with confidence and precision, both AC and DC.

change of the output error that is of interest, not really the p-p value before or after, but the delta. You do not *have to* trim the resistor to get the slope perfect; but that is the easy way for the guy working at his bench to see the changes.

This is a great circuit to fool around with. When you get it running, you will want to test every op amp in your area, because it gives you such a neat high-resolution view. It gives you a good *feel* for what is happening, rather than just hard, cold, dumb numbers. For example, if you see a 22-mV p-p output signal that is caused by a 22-µV error signal, you know that the CMRR really is way up near a million, which is a lot more educational than a cold "119.2 dB" statement. Besides, you learn rather quickly that the slope and the curvature of the display are important. Not all amplifiers with the same "119.2 dB" of CMRR are actually the same, not at all. Some have a positive slope, some may have a negative slope, and some curve madly, so that if you took a two-point measurement, the slope would change wildly, depending on which two points you choose. (If you increase the amplitude of the input signal, you can also see plainly where severe distortion sets in—that's the extent of the common-mode range.)

Limitations: If you set the noise gain as high as 100, then this circuit will be 3 dB down at F_{GBW} divided by 100, so you would only use this up to about 1 kHz on an ordinary 1 MHz op amp, and only up to 100 Hz at a gain of 1000. That's not too bad, really.

To look at CMRR above 1 kHz, you might use R100c = 2 k, to give good results up to 10 kHz. In other words, you have to engineer this circuit a little, to know where it gives valid data. Thinking is required. Sorry about that.

For really fast work, I go to a high-speed low-gain version where R1 = R11 = 5 k, R2 = R12 = 5 k, and R100 = 2 k or 1 k or 0.5 k. This works pretty well up to 50 kHz or more, depending on what gain-bandwidth product your amplifier has.

For best results at AC, it's important to avoid stray capacitance of wires or of a real switch at the points where you connect to R100a or R100b. Usually I get excellent results from just grabbing on to the resistor with a minigator clip. You can avoid the stray pf that way; if you use a good selector switch, with all the wires dressed neatly in the *air* (which is an excellent insulator) you may be able to get decent bandwidth, but you should be aware that you are probably measuring the AC CMRR of your set-up, not of the op amp.

I was discussing this circuit with a colleague, and I realized the right way to make this R100 is to solder, for example, 100 Ω to the + input and 100 Ω to the – input, and then just clip their tips together to make 200 Ω—balanced strays, and all that.

If you have an op amp with low gain or low g_m, you may want to add in a buffer follower at a-b, so the amplifier does not generate a big error due to its low gain. The National LM6361 would need a buffer as it only has a gain of 3000 with a load of 10 kΩ, and its CMRR is a lot higher than 3000.

Altogether, I find this circuit has better resolution and gives less trouble than any other circuit for measuring CM error. And the price is right: a few resistors and a minigator clip.

Single-Supply Operation??

One of the biggest applications problems we have is that of customer confusion about single-supply operation. Every week or so, we get a call from a customer: "Can I run your LM108 (or LF356 or LM4250) on a single power supply? Your data sheet doesn't say whether I can. . . ."

Sigh. We cheerfully and dutifully explain that you can run *any* operational amplifier on a single supply. An LM108 does not have any "ground" pin. It can't tell if your power supplies are labelled "+15 V and –15V" or "+30 V and ground," or "ground and –30 V." That is merely a matter of nomenclature—a matter of your standpoint—a matter of which bus you might choose to call ground.

Now it is true that you have to bias your signals and amplifier inputs in a reasonable way. The LM108/LM308 can amplify signals that are not too close to either supply—the inputs must be at least a couple volts from either supply rail. So if you need a circuit that can handle inputs near the minus rail, the LM108 is not suitable, but the LM324, LM358, LMC660, LMC662, and LM10 *are* suitable. If you need an amplifier that works near the + rail, the LM101A and LM107 are guaranteed to work there. (The LF355 and LF356 typically work well up there but are not guaranteed.)

But if you keep the inputs biassed about halfway between the rails, just about any op amp will work "on a single supply." It's just a matter of labels! Someday we plan to write an applications note to get these silly questions off our back, and to answer the customers' reasonable questions, but right now everybody's a little too busy to write it.

Measure Bias Current Rather Than Impedance

Another op amp spec you don't need to worry about is the differential input impedance. Every year I still get asked, "How do we measure the input impedance of an op amp?" And every year I trot out the same answer: "We don't." Instead, we measure the bias current. There's a close correlation between the bias current and the input impedance of most op amps, so if the bias current is low enough, the input impedance (differential and common mode) must be high enough. So, let's not even think about how to measure the low-frequency differential input impedance (or, input resistance) because I haven't measured it in the last seven years.

Generally, an ordinary differential bipolar stage has a differential input impedance of $1/(20 \times I_b)$, where I_b is the bias current. But this number varies if the op amp includes emitter-degeneration resistors or internal bias-compensation circuitry. You can easily test the common-mode input resistance by measuring I_b as a function of V_{CM}.

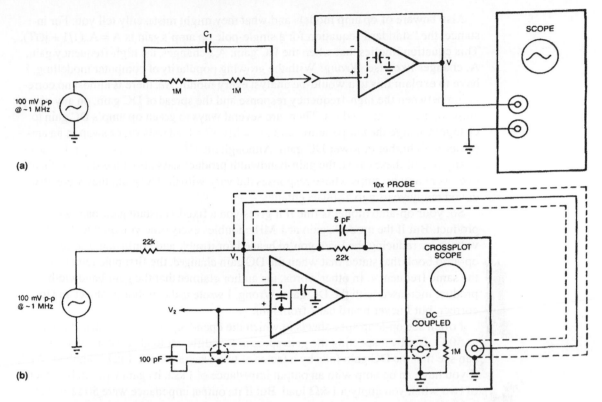

Figure 8.8. Circuit (a) lets you test an op amp's common-mode input capacitance. When $C_1 = C_A \approx 5$ pF, measure $V_1 = V_A$; when $C_1 = C_B \approx 1000$ pF, measure $V_1 = V_B$. Then $C_{in} = C_A \times (V_B - V_A)/[V_A - V_B \times C_A/C_B]$. For best results, connect the signal to the plus input of the DUT with a small gator clip. Do not put the plus input pin into the device's socket. Circuit (b) lets you test an op amp's differential input capacitance. for this circuit, $C_{in\,(differential)} = V_{2(p-p)} \times C_{total}/[V_{1(p-p)} - V_{2(p-p)}]$, where $C_{total} = C_{in(common\,mode)} + C_{cable} + C_{scope\,in} + 100$ pF.

I've measured some input capacitances and find the circuits of Figure 8.8 to be quite useful. Input-capacitance data is nominally of interest only for high-impedance high-speed buffers or for filters where you want to make sure that the second-source device has the same capacitance as the op amps that are already working okay.

Recognize False "Error" Characteristics

Sometimes, an op amp may exhibit an "error" that looks like a bad problem, but isn't. For example, if you have your op amp's output ramping at –0.3 V/μsec, you might be surprised when you discover that the inverting input, a summing point, is *not* at ground. Instead, it may be 15 or 30 or 100 mV away from ground. How can the offset voltage be so bad if the spec is only 2 or 4 mV?

Why isn't the inverting input at the "virtual ground" that the books teach us? The virtual ground theory is applicable at DC and low frequencies, but if the output is moving at a moderate or fast speed, then expecting the summing point to be exactly at ground is unreasonable. In this example, dV_{out}/dt equals 2π times the unity-gain frequency times the input voltage. *So*, 15 mV of V_{in} is quite reasonable for a medium-bandwidth op amp, such as an LF356, and 50 or 70 mV is quite reasonable for an LM741. If you want an op amp to move its output at any significant speed, there has to be a significant error voltage across the inputs for at least a short time.

Also beware of op amp models and what they might mistakenly tell you. For instance, the "standard" equation for a single-pole op amp's gain is $A = A_o(1/1 + j\omega T)$. This equation implies that when the DC gain, A_o, changes, the high-frequency gain, A, changes likewise. *Wrong!* With the growing popularity of computer modelling, I have to explain this to a would-be analyst every month. *No*, there is almost no correlation between the high-frequency response and the spread of DC gain, on any op amp you can buy these days. There are several ways to get an op amp's DC gain to change: Change the temperature, add on or lift off a load resistor, or swap in an amplifier with higher or lower DC gain. Although the DC gain can vary several octaves in any one of these cases, the gain-bandwidth product stays about the same. If there ever were any op amps whose responses did vary with the DC gain, they were abandoned many years ago as unacceptable.

So, your op-amp model is fine if it gives you a fixed, constant gain-bandwidth product. But if the model's gain at 1 MHz doubles every time you double the DC voltage by reducing the load, you're headed for trouble and confusion. I once read an op-amp book that stated that when the DC gain changed, the first pole remained at the same frequency. In other words, the author claimed that the gain-bandwidth product increased with the DC gain. Wrong. I wrote to the author to object and to correct, but I never heard back from him.

I often see op-amp spec sheets in which the open-loop output impedance is listed as 50 Ω. But by inspecting the gain specs at two different load-resistor values, you can see that the DC gain falls by a factor of two when a load of 1 kΩ is applied. Well, if you have an op amp with an output impedance of 1 kΩ, its gain will fall by a factor of two when you apply a 1-kΩ load. But if its output impedance were 50 Ω, as the spec sheet claimed, the gain would only fall 5%. So, whether it's a computer model or a real amplifier, be suspicious of output impedances that are claimed to be unrealistically low.

Watch Out for Real Trouble

What real trouble can an op amp get you into without much help from the components? Well, you could have a part with a bad V_{OS}. Or if the temperature is changing, the thermocouples of the op amp's Kovar leads may cause small voltage differences between the op amp leads and the copper of the PC board. Such differences can amount to 1/10 or 1/20 of a Celsius degree times 35 μV per degree, which equals 2 to 3 μV. A good way to avoid this problem is to put a little box over the amplifier to keep breezes and sunshine off the part. That's very helpful unless it's a high-power op amp. Then just repeat after me, "Heat is the enemy of precision," because it is. After all, when an IC has to handle and dissipate a lot of power, it's not going to be nearly as accurate as when it is not overheating, and when all the components around it get heated, too.

You should remember, too, that not all op amps of any one type have the exact same output-voltage swing or current drive or frequency response. I get phone calls every four or five months from customers who complain, "We have a new batch of your op amps, and they don't have as good an output swing (or output current or frequency response) as the previous batches." When I check it out, 98% of the time I find that a part with extremely good performance is just a random variation. The customer had got into the habit of expecting *all* the parts to be better than average. When they got some parts that were still much better than the guaranteed spec but worse than average or "typical," they found themselves in trouble. It's always painful to have to tell your friends that you love them when they like you and trust you, but

Pease's Principle

For many years, I've been cautioning people: If you have a regulator or an amplifier circuit and it oscillates, *do not* just add resistors or capacitors until the oscillation stops. If you do, the oscillation may go away for a while, but after it lulls you into complacency, it will come back like the proverbial alligator and chomp on your ankle and cause you a great deal of misery.

Instead, when you think you have designed and installed a good fix for the oscillation, BANG on the output with square waves of various amplitudes, frequencies, and amounts of load current. One of the easiest ways to perform this test is to connect a square-wave generator to your circuit through a couple hundred ohms in series with a 0.2-µF ceramic disc capacitor. Connect the generator to the scope, so you can trigger on the square-wave signal. Also, apply an adjustable DC load that's capable of exercising the device's output over its full rated output-current range or, in the case of an op amp, over its entire rated range for output voltage *and* current.

To test an op amp, try various capacitive loads to make sure it can drive the worst case you expect it to encounter. For some emitter-follower output stages, the worst case may be around 10 to 50 pF. The

oscillation may disappear with heavier capacitive loads.

The adjacent figure does not show a recommended value for some of the parts. If you are testing a 5-A regulator, you may want a load resistor as low as an ohm or two. If you are evaluating a low-power amplifier or a micropower reference, a value of a megohm may be reasonable, both for the load resistor and for the resistor from the square-wave generator. Thinking is recommended. Heck, thinking is *required*.

When you *bang* a device's output and that output rings with a high Q, you know your "fix" doesn't have much margin. When the output just goes "flump" and doesn't even ring or overshoot appreciably, you know your damping is effective and has a large safety margin. Good! Now, take a hair dryer and get the circuit good and warm. Make sure that the damping is still pretty well behaved and that the output doesn't begin to ring or sing when you heat the capacitor or power transistor or control IC or *anything*.

I don't mean to imply that you shouldn't do a full analysis of AC loop stability. But the approach I have outlined here can give you pretty good confidence in about five minutes that your circuit will (or won't) pass a full set of exhaustive tests.

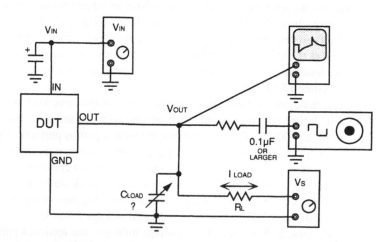

they really shouldn't trust your parts to always be better than average. Maybe in Lake Wobegon, all the kids are better than average, but you can't go shopping for op amps and complain when they are not all "better than average."

Oscillations Do Occasionally Accompany Op Amps

One of the most troublesome problems you can have with op amps is oscillation. Just as you can build an oscillator out of any gain block, then you must admit that any gain block can also oscillate when you don't want it to. Op amps are no exception. Fortunately, most op amps these days are well behaved, and you only need to take four basic precautions to avoid oscillations.

First, always use some power-supply bypass capacitors on each supply and install them near the op amp. For high-frequency op amps, the bypass capacitors should be very close to the device for best results. In high-frequency designs, you often need ceramic *and* tantalum bypass capacitors. Using bypass capacitors isn't just a rule of thumb, but a matter of good engineering and optimization.

Second, avoid unnecessary capacitive loads; they can cause an op amp to develop additional phase shift, which makes the op-amp circuit ring or oscillate. These effects are especially noticeable when you connect a $1\times$ scope probe or add a coaxial cable or other shielded wire to an op amp, to convey its output to another circuit. Such connections can add a lot of capacitance to the output. Unless you're able to prove that the op amp will be stable driving that load, you'd better add some stabilizing circuits. It doesn't take a lot of work to *bang* the op amp with a square wave or a pulse and see if its output rings badly or not. You should check the op amp's response with both positive and negative output voltages because many op amps with pnp-follower outputs are less stable when V_{out} is negative or the output is sinking current. Refer to the box, "Pease's Principle."

I've seen pages of analysis that claim to predict capacitive-loading effects when the op amp's output is resistive, but as far as I am concerned, they're a complete waste of time: The output impedance of an op amp is usually *not* purely resistive. And if the impedance is low at audio frequencies, it often starts to rise inductively at high frequencies, just when you need it low. Conversely, some op amps (such as the NSC LM6361) have a high output impedance at low frequency, which falls at high frequencies—a capacitive output characteristic, so when you add more capacitance on the output, the op amp just slows down a little and doesn't change its phase very much. But if an op amp is driving a remote, low-resistive load that has the same impedance as the cable, the terminated cable will look resistive at all frequencies and capacitive loading may not be a problem. (But you still have to be able to drive that low-impedance 75-Ω load!)

You can decouple an inverter's and integrator's capacitive load as shown in Figure 8.9. If you choose the components well, any op amp can drive any capacitive load from 100 pF to 100 µF. The DC and low-frequency gain is perfectly controlled, but when the load capacitor gets big, the op amp will slow down and will eventually just have trouble slewing the heavy load. Good starting-point component values are R1 = 47 to 470 Ω and CF = 100 pF. These values usually work well for capacitive loads from 100 pF to 20,000 pF. If you have to make an integrator or a follower, you'll need an additional 4.7 kΩ resistor as indicated in Figure 8.9(d).

In some cases, as with an LM110 voltage follower, the feedback path from the output to the inverting input is internally connected and thus unavailable for tailoring. In this case, we can pull another trick out of our bag: The tailoring of noise gain. Noise gain is defined as $1/\beta$, where β is the attenuation of an op amp's feedback network, as seen at the op amp's inputs, referred to the output signal. For instance, the β of the standard inverter configuration (Figure 8.10a) equals $Z_1/(Z_1 + Z_2)$, so the noise gain equals N + 1. You can raise the noise gain as shown in Figure 8.10b.

If you're using a low-noise-gain op-amp configuration (such as a unity-gain follower that has a noise gain of 1 (Figure 8.11a)), it's well known that for good stability, the op amp and its feedback network can't have appreciable unwanted phase shift out near its unity-gain frequency. If you can increase the noise gain to 4 or 5, the requirement for low phase shift eases considerably. No, you don't have to change the signal gain to 5. A noise gain of 5 or greater is easy to achieve (Figure 8.11b) while maintaining a gain of 1 for the signal. Even the unity-gain follower with a wire from the output to the inverting input can be saved, as illustrated in Figures 8.11c and d. You'll find a more complete description of these circuits in Ref. 2, which I wrote in 1979, but meanwhile, if you are having stability problems with followers, just go ahead and try these techniques—it's as easy as adding a resistor box or a pot to your existing circuit. I should also mention that some of these concepts were used by

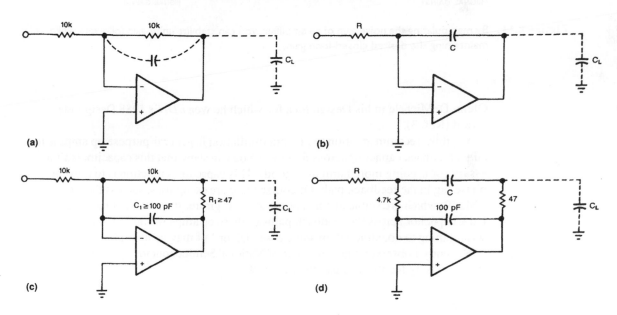

Figure 8.9. You can easily modify the basic inverter (a) and integrator (b) to decouple capacitive loads (c) and (d).

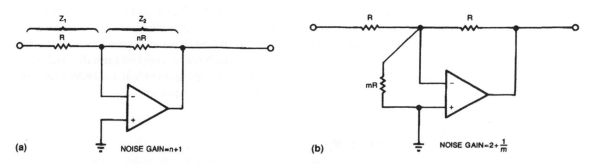

Figure 8.10. By adding just a single resistor, you can tailor the noise gain of a standard integrator.

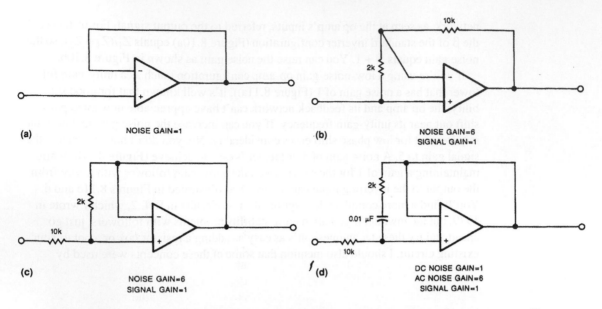

Figure 8.11. By manipulating the noise gain of an amplifier, you can stabilize unity-gain followers while maintaining the desired closed-loop gain.

Glenn DeMichele in his Design Idea for which he won *EDN*'s 1988 Design Idea award (Ref. 3).

My third recommendation to prevent oscillation in general-purpose op amps is to add a feedback capacitor across R_F unless you can show that this capacitor isn't necessary (or is doing more harm than good). This capacitor's function is to prevent phase lag in the feedback path. Of course there are exceptions, such as the LF357 or LM349, which are stable at gains or noise gains greater than 10. Adding a big feedback capacitor across the feedback paths of these op amps would be exactly the wrong thing to do, although in some cases 1/2 or 1 pf may be helpful....

Recently I observed that a number of National Semiconductor op-amp data sheets were advising feedback capacitor values of

$$C_F = \frac{C_{in} R_{in}}{R_F}$$

But, if you had an ordinary op amp whose C_{in} was 5 pF and an inverter with a gain of −0.1, with $R_F = 1$ MΩ and $R_{in} = 10$ MΩ, this equation would tell you to use a C_F of 50 pF and accept a frequency response of 3 kHz. That would be absurd! If you actually build this circuit, you'll find that it works well with $C_F = 1.5$ pF, which gives the inverter a bandwidth of 100 kHz. So, we at NSC have just agreed to deep-six that equation. We have a couple new formulas, which we've checked carefully, and we have found that you can get considerably improved bandwidth and excellent stability. For high values of gain and of R_F, use the following equation:

$$C_F = \sqrt{\frac{C_{in}}{GBW \times R_F}}$$

where GBW is the gain-bandwidth product. In those cases in which the gain or impedance is low, such as where

$$\left(1 + R_F/R_{in}\right) \leq 2\sqrt{GBW \times R_F \times C_{in}}$$

use the following equation

$$C_F = \frac{C_{in}}{\left(2 + 2R_F/R_{in}\right)}$$

I won't bore you with the math, but these equations did come from real analytical approaches that have been around for 20 years—I championed them back at Philbrick Researches. The value of C_F that you compute is not that critical; it's just a starting point. You really must build and trim and test the circuit for overshoot, ringing, and freedom from oscillation. If the equation said 1 pF and you get a clean response only with 10 pF, you'd be suspicious of the formula. Note that when you go from a breadboard to a PC board, the stray capacitances can change, so you must recheck the value of C_F. In some cases, you may not need a separate capacitor if you build 0.5 pF into the board. In any case, you certainly don't have to just trust my equations—build up your breadboard and fool around with different values of C_F and check it out for yourself. See if you don't agree.

My last recommendation is that when you think the circuit is okay—that is, free from oscillation—test it anyway per Pease's Principle to make sure it's as fast or stable as you expect. Be sure that your circuit isn't ringing or oscillating at any expected operating condition or load or bias.

Noises, Theoretical and Otherwise

In addition to oscillatory behavior, another problem you might have when using op amps is noise. Most op amps have fairly predictable noise. It's often right down at theoretical levels, especially at audio frequencies. There's a pretty good treatment of noise and its effects in various applications in Thomas Frederiksen's book (Ref. 4). Also, if you want to optimize the noise for any given source resistance or impedance, National Semiconductor's Linear Applications Note AN222 (Ref. 5) has some good advice, as does the article in Ref. 6.

You'll have difficulty with noise when it's unpredictable or when op amps of a particular type have varying noise characteristics. This problem rarely happens at audio frequencies but is likely to happen sporadically at low frequencies, such as 10 or 100 Hz or even lower frequencies. Every manufacturer of transistors and amplifiers tries to keep the noise low, but occasionally some noisy parts are built. Sometimes the manufacturer is able to add tests that screen out the noisy parts. But these tests aren't cheap if they take even one second of tester time, which may cost three cents or more. We are trying to add some 0.3-second tests to some of our more popular amplifiers, but it's not trivially easy.

Here's a tip—we find that true RMS testing for noise is a big waste of time, because amplifiers that are objectionably noisy have much worse p-p noise than you would guess from their RMS noise data. So the p-p test is the best discriminator. We get the best resolution with a bandwidth from around 30 Hz to 3 kHz, after we roll off the broadband noise. A sample time of 0.1 second works pretty well; the ones that pass a 0.1-second test but fail a 0.5-second test are uncommon.

Popcorn Noise Can Rattle Sensitive Circuits

Flicker noise, also known as 1/f noise, is AC noise that exists at low frequencies. And even more insidious than 1/f noise is popcorn noise—a type of electrical noise in which bursts of square steps are added to the normal thermal noise at random times. Popcorn noise occurs rarely these days, but, unfortunately, it's not at 0%, not even with the cleanest processing and the best manufacturers. I've been chastised and told that some of my amplifiers are noisy compared with those of certain competitors. But when I look at the competitor's data and plots, I see 1/f and popcorn noise lurking unnoticed in a corner. On high-performance parts, we try to screen out the noisy ones. But when a few parts have a spacing of 2 to 10 seconds between bursts of popcorn, it's not cost-effective to look for those parts. Only a small percentage of our customers would want to reject that one noisy part *and* pay for the testing of all the good parts, too. Remember, 10 seconds of testing time equals 30 cents; time equals money.

Although oscillation and noise problems may be the most common ones you'll encounter when you use op amps, there's a host of other characteristics that are wise to look out for. These characteristics include overload or short-circuit recovery, settling time, and thermal response. Many op amps have a fairly prompt recovery from overdrive when you make the output go into the stops—that is, when you force the output into the power-supply rails. For most op amps, this recovery characteristic is not defined or specified. One recently advertised op amp claims to require only 12 ns to come out of the stops. Just about all other op amps are slower to one degree or another. The recovery time for chopper-stabilized amplifiers can be *seconds*.

Even if you have a fast op amp that doesn't have a delay coming out of limit, there may be circuits, such as integrators, that take a long time to recover if you overdrive the output and inputs. To avoid these cases, a feedback bound made of zeners and other diodes may be helpful (Ref. 7). However, if you have a differential amplifier, you may not be able to use any zener diode feedback limiters. I recall the time I designed a detector circuit using a fast, dielectrically isolated op amp. When I went to put it into production, nothing worked right. It turned out that the manufacturer had just recently redesigned the chip to cut the die size by 50%. The new and improved layout *just happened to* slow down the op amp's overdrive-recovery time. I wound up redesigning the circuit to use an LM709. I saved a lot of pennies in the long run, but the need to change parts didn't make me very happy at the time.

Rely Only on Guaranteed Specs

Don't rely on characteristics that aren't specified or guaranteed by the manufacturer. It's perfectly possible for you to test a set of samples and find that they feature some desired performance characteristic that is not specified by the manufacturer. But if the next batch doesn't fulfill your requirements, whom are you going to get angry at? Don't get mad at me, because I'm warning you now. Any unspecified conditions may cause a test result to vary considerably compared to a guaranteed tested specification. If you have to work in an unspecified range, you should keep a store of tested good ICs in a safe as insurance. If a new batch comes in and tests "bad" you'll have some backup devices. I recall a complaint from a user of LM3046 transistor arrays: A fraction of the parts failed to log accurately over a wide range. The "bad" ones turned out to have a beta of 20 at a collector current of 50 pA, versus a beta of 100 for the "good" ones. I convinced the user that keeping a few hundred of these inexpensive parts in a safe (yes, literally) would be a lot cheaper than getting the manufacturer to sort out high-β devices.

Op amps and other linear ICs can also have errors due to thermal "tails." These tails occur when the change of heat in one output transistor causes a thermal gradient to sweep across the chip. This change occurs gradually, often over milliseconds, and causes uneven heating of input transistors or other sensitive circuits. Many high-power circuits and precision circuits, such as the LM317, LM350, LM338, LM396, LM333, and LM337, have been tested for many years for thermally caused error. These tests aren't performed only on power ICs, but also on precision references such as LM368 and LM369, and on instrument-grade op amps. In fact, a recent article by a Tektronix engineer (Ref. 9) points out that thermal tails can be a major source of error in fast signal amplifiers and that innovative circuit design can minimize those overdrive-recovery errors.

If you have ever studied the gain errors of older OP-07-type amplifiers, you have probably recognized that these errors and nonlinearities were caused by thermal errors—which were related to a bad layout. These days, most OP-07s have a better layout, and the thermal distortions have been banished.

Another characteristic that is not specified or discussed is the change of offset voltage vs. stress. This is most noticeable on BIFET amplifiers, as the FETs are much more sensitive to stresses in the silicon die than bipolar transistors are. When you install and solder a plastic DIP op amp in a PC board, and then warp the board, you can monitor the V_{OS} and watch it shift. Some amplifiers are better than others of similar types. It has a lot to do with the layout and also with the die attach. If you need the lowest offset, watch out for this. If the board is vibrated, the AC warping and stress can cause microphonic AC noises, too. CER-DIP amplifiers have a stronger ceramic base, and have a little less of a problem.

Here's where it gets really wild: Buy your BIFET amplifiers in an SO (Small-Outline, surface-mount) package. The smaller plastic package is able to take up even less of the stress, and the die gets warped even more, and the change of V_{OS} gets even worse than in ordinary DIPs. So, when you think you can pack even more of a good thing onto a board by going to surface-mount (Small-Outline) packages, you may also pack in more trouble. There is no specification on this on any data sheet. So a SPICE analysis has no way to warn you about this potential problem. Even a breadboard does not necessarily tell you about this. The actual prototype units, on the real PC boards, must be checked out.

These days, just about every manufacturer's monolithic op amps will survive a short from the output to ground. (Hybrids are often unprotected.) But it's not always clear whether an op amp will survive a short circuit to the positive or negative supply or, if so, for how long. You may have to ask the manufacturer, and you can expect some kind of negative answer. You'll be told to avoid overheating the device above its absolute maximum junction temperature. Even if an amplifier or regulator does recover fairly quickly from current limit, nobody will guarantee that it won't oscillate when in current limit. Nor will the manufacturer have much knowledge about how the circuit recovers from the thermal gradients caused by current limit. If an op amp survives a high-power overload, it's not fair to ask the device to recover its full accuracy very quickly. The most you really can ask for is that it survives with no degradation of reliability—that's the standard.

Some op amps (such as LM12 and LM10) and most voltage regulators (and other power ICs) have an on-chip temperature limiter. Thermal shutdown circuits can improve reliability. If a heavy overload is applied for a long time, or there is no heat sink, or the ambient is just too hot, these circuits detect when the chip's temperature exceeds 150 °C and then turn off the output. The thermal-limiter circuit in the LM117 (and other early power ICs) sometimes just decreased the output current to a safe DC value to hold the die temperature to around 160 °C. In other cases, where the load is lighter and the thermal gradient transients are different, these thermal limiters oscillate ON and OFF with a duty cycle that ensures the 160 °C chip temperature. As I was about to design the LM137, I looked back and decided the latter characteristic was preferable, so I designed about 5 °C of thermal hysteresis into the thermal-limit circuit. That way, the circuit makes a strong attempt to restart its heavy load with a repetition rate of about 100 Hz. If the regulator makes only a feeble attempt, it may be unable to start some legal loads.

So, we actually designed an oscillation into this thermal-limit circuit, but we never bothered to mention it on the data sheet. H'mmm... we shouldn't be so sloppy. I apologize. I'll do better next time. (This situation has a bearing on one of my pet peeves: Bad data sheets. I get really cross about bad ones, and I really do try hard and work hard to make good ones. Refer to "How to Read a Data Sheet" (Appendix F and Ref. 10) because bad data sheets can get the user into trouble.)

Different Methods Uncover Different Errors

Now that you know some op-amp problems to look out for, how do you actually troubleshoot an op-amp circuit? I usually split my plan along two lines: AC and DC problems. Examples of AC problems include oscillations and ringing; DC problems include bad DC output errors and pegged outputs, which are outputs stuck at either the positive or negative supply rail. Obviously, you need a scope to be sure the circuit isn't oscillating. It always makes me nervous when I find out that the customer I'm trying to help *doesn't even have a scope*. I can understand if an engineer only has a crummy scope, but there are certain problems you cannot expect to solve—nor can you even verify a design—if you don't have any scope at all.

If the problem is an AC problem, I first make sure that the input signals are well behaved and at the values I expect them to be. Then I put my scope probes on all the pins and nodes of the circuit. Sometimes it's appropriate to use a 10 × probe, and other times I shift to 1 × mode. Sometimes I AC-couple the scope; sometimes I DC-couple it. I check all the pins, especially the power-supply pins. Then—depending on what clues I see—I poke around and gather symptoms by adding capacitors or RC boxes to assorted circuit nodes. I try to use two probes to see if the input and output have an interesting phase relationship, and I simultaneously verify that the output is still oscillating.

Many of the techniques I use depend on whether the circuit is one I've never tried before or one that I see all the time. Sometimes I find an unbelievable situation, and I make sure that I understand what's going on before I just squash the problem and proceed to the next. After all, if I'm fooling myself, I really ought to find out how or why, so I won't do it again.

If the op amp exhibits a DC error or a peg, I first check with my scope to see that there's no oscillation. Then I bring in my 5-digit DVM and scribble down a voltage map on a copy of the schematic. On the first pass, I'm likely to just keep the numbers in my head to see if I can do a quick diagnosis of a problem that's obvious, such as a bad power supply, or a ground wire that fell off, or a missing resistor. Failing that, I start writing meticulous notes to help look for a more insidious problem. I look at the numbers on the schematic and try to guess the problem. What failure could cause that set of errors? A resistor of the wrong value? A short? An open? Then I try to cook up a test to confirm my theory. Sometimes I have to disconnect things, but I try to minimize that. Sometimes adding a resistor or voltage or current will yield the same result, and it's much easier than disconnecting components.

If an amplifier circuit isn't running at all, sometimes the right thing to do is to reach into the circuit and "grab" one amplifier's input and force it to go above and below the other input. If the output doesn't respond at all, you have a dead amplifier, an amplifier with no connections, or a stuck output. It is not obvious to try this open-loop test—no book tells you that this is a good idea—but after you try it, you will agree that its results usually tell you an obvious story. Refer to Figure 14.1 in Chapter 14 for more detailed techniques and notes on troubleshooting basic op-amp circuits.

Many of these op-amp troubleshooting tips are applicable to other components as well. The next chapter will continue with buffers, comparators, and related devices.

References

1. *Data Converter Handbook*, Analog Devices, P.O. Box 9106, Norwood MA 02062, 1974.

2. Pease, Robert A., "Improved unity-gain follower delivers fast, stable response," *EDN*, February 20, 1979, p. 93. (Also available as LB-42 in NSC's Linear Applications Book, 1980, 1986, 1989, etc., "Get Fast Stable Response from Improved Unity-Gain Followers.")

3. DeMichele, Glenn, "Compensate op amps without capacitors," *EDN*, July 21, 1988, p. 331.

4. Frederiksen, Thomas M., *Intuitive Operational Amplifiers*, McGraw-Hill, New York, NY, 1985. Available from Heath Company, P.O. Box 8589, Benton Harbor, MI 49022. (800) 253-0570 (Part No. EBM-1), $19.95.

5. Nelson, Carl T., *Super Matched Bipolar Transistor Pair Sets New Standards for Drift and Noise*, Application Note AN-222, Linear Applications Databook, p. 517, National Semiconductor, Santa Clara, CA, 1986.

6. Pease, Robert A., "Low-noise composite amp beats monolithics," *EDN*, May 5, 1980, p. 179. (Also available as LB-52 in NSC's Linear Applications Databook, 1982, 1986, 1989, etc. as "A Low-Noise Precision Op Amp.")

7. Pease, Robert A., "Bounding, clamping techniques improve on performance," *EDN*, November 10, 1983, p. 277.

8. Pease, Bob, and Ed Maddox, "The Subtleties of Settling Time," *The New Lightning Empiricist*, Teledyne Philbrick, Dedham, MA, June 1971.

9. Addis, John, "Versatile Broadband Analog IC," *VLSI Systems Design*, September 1988, p. 18.

10. Pease, Robert A., "How To Get The Right Information From A Datasheet," *EE Times*, April 29, 1985, p. 31. (Also available as Appendix F in NSC's General-Purpose Linear Devices Databook, 1988, 1989, etc. and as Appendix F in this book.)

in real-time step response. If that is consistent with the frequency-domain response, fine; if not, I get suspicious. . . .)

Secondly, if an earlier version of your circuit has worked OK, what's the difference between the new one that does not work well and the old one that does? Be sure to keep one or more examples of the old version around so that you can make comparisons when the new circuits have troubles. (Note that I said *when*, not *if*.) Thirdly, look for components such as capacitors whose high-frequency characteristics can change if someone switched types or suppliers.

Optoisolators in switching regulators are another possible cause of oscillation trouble due to their wide range of DC gain and AC response. A switching-regulator IC, on the other hand, is not as likely to cause oscillations, because its response would normally be faster than the loop's frequency. But, the IC is never absolved until proven blameless. For this reason, you should have an extra module with sockets installed just for evaluating these funny little problems with differing suppliers, variant device types, and marginal ICs. You might think that the sockets' stray capacitances and inductances would do more harm than good, but in practice, you can learn more than you lose.

When Is an Oscillation Not an Oscillation?

We still get a phone call, every month or so, from somebody complaining about a "120-Hz oscillation" on one of our circuits. (It's a good thing we do, because one of our applications engineers was mentioning such a case recently, and I realized I had forgotten to mention this type of oscillation, so this paragraph got plopped into the text at the last minute. If I hadn't remembered to include this class of "oscillation," I would have been terribly embarrassed.) Now, how can an op amp be "oscillating" at 60 or 120 Hz? Well, it is not *impossible* for an op amp or regulator to oscillate at this frequency, but it is *extremely* unlikely. What is surely happening is that there is some noise at power-line frequency getting into the circuit. There are four major ways for this to happen.

1. When there is a diode connected to a delicate input, the ambient light in the room can shine in and generate photocurrents. With fluorescent lights, this is usually at 120 Hz, but the higher harmonics can boom in, too, along with some DC current. As soon as you realize this is happening, it's fairly easy to troubleshoot this by adding a little darkness to the circuit—cover it up with a dark cloth, jacket, or book. You can then localize it and "darken" it permanently. In the case of extremely bright light, it can even come in through the insulators in the base of a T0-99 can—the little ceramic feed-throughs are not really opaque—they let a small fraction of the light in. Fortunately, plastic DIP packages are very opaque, these days.

2. A power supply can have more 60- or 120-Hz ripple (saw-tooth shape or pulse-shape) than you expect. This can be caused by bad connections, a bad capacitor, an open rectifier, or a ground loop. Again, as soon as you recognize that this kind of thing *can* happen, it's easy to search and cure the problem.

3. Magnetic flux from a transformer gets coupled into your circuit. The two most common sources are a soldering iron close by, or, a power transformer that is saturating a little bit, spraying flux around. This usually has a distinctive shape at 60 Hz with lots of harmonics, and is quite position sensitive. This, too, is fairly easy to recognize. But if you discover that your power transformer is not only running hot, but spraying flux badly, it's not usually easy to relocate it if you are nearly done with your project. I would love to recommend that you assemble your power supply in a little box, 3 feet away from your main instrument, but that is not always feasible. You

should at least start out by installing only a power transformer of known quality, with good, known freedom from excessive external flux and saturation. Sometimes you can retrofit your circuit with a toroidal power transformer, but most people don't keep these lying around. They are more expensive but often worth it, as they are more efficient and have less self-heating.

In concept, you might try adding some shielding. Go ahead, put in 1/16 inch of aluminum. It will have no effect at all—for magnetic shielding, you need iron. Go ahead, put in 1/16 inch of iron. That's not much help, because at 60 Hz, it takes about 1/4 inch thickness of steel to do much good. You can try, but it's not an easy way. Sometimes you can arrange your critical circuits to have smaller loops, so they will pick up less flux. Make neat, compact paths; use twisted pairs; and use layout tricks like that—those can sometimes help. If you haven't tried these before, ask an old-timer.

4. A mechanical vibration can be coupled in through a floppy wire or a high-K ceramic capacitor. If nobody tells you about this one, this is a very tricky tease, not at all easy to guess. Sometimes if you replace the high-K ceramic with an NPO, or a film capacitor, it will solve the problem. Recently we ran a picoammeter, and when the power supply lead ran near the summing point, there was a certain amount of charge, Q = C × V. When the wire was vibrated at line frequency, a 60-Hz current I = V × dC/dT flowed into the input. The current stopped when we guarded the 5-V bus away from the input, and we also added shock-mounting for the whole assembly, to keep out all vibrations.

There are probably a few other ways to get 60-Hz noises into a circuit, so you must be prepared to exercise ingenuity to search for nasty coupling modes. But if the "oscillation" is at exactly line frequency, and it synchronizes with the "line synch" mode of your scope, then it is certainly not a real oscillation. Now, I *have* seen 59-Hz oscillations, that would fool you into thinking they were at 60 Hz, but that is quite rare. It just goes to show that there are many noises to keep you on your toes. Some are oscillations, and some are "oscillations."

You can best analyze the design of a slow servo mechanism, such as that in Figure 9.2, with a strip-chart recorder because the response of the loop is so slow. (A storage scope might be OK, but a strip-chart recorder works better for me.) You might wish to analyze such a servo loop with a computer simulation, such as SPICE, but the thermal response from the heater to the temperature sensor is strictly a function of the mechanical and thermal mounting of those components. This relationship would hardly be amenable to computer modelling or analysis.

Comparators Can Misbehave

Saying that a comparator is just an op amp with all the damping capacitors left out— that is an oversimplification. Comparators have a lot of voltage gain and quite a bit of phase shift at high frequencies; hence, oscillation is always a possibility. In fact, most comparator problems involve oscillation.

Slow comparators, such as the familiar LM339, are fairly well behaved. And if you design a PC-board layout so that the comparator's outputs and all other large, fast, noisy signals are kept away from the comparator's inputs, you can often get a good clean output without oscillation. However, even at slow speeds, an LM339 can oscillate if you impress a slowly shifting voltage ramp on its differential inputs. Things can get even messier if the input signals' sources have a high impedance ($\gg$10 kΩ) or if the PC-board layout doesn't provide guarding.

In general, then, for every comparator application, you should provide a little hysteresis, or positive feedback, from the output back to the positive input. How

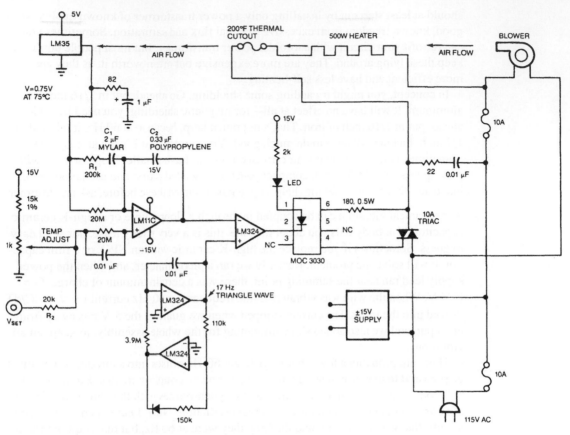

Figure 9.2. Stabilizing this heater's slow servo loop and choosing the proper values for R_1, C_1, and C_2 involved applying a 1-V p-p, 0.004-Hz square wave, V_{SET}, to R_2 and observing the LM11C's output with a strip-chart recorder.

much? Well, I like to provide about twice or three times as much hysteresis as the minimum amount it takes to prevent oscillation near the comparator's zero-crossing threshold. This excess amount of feedback defines a safety margin. (For more information on safety margins, see the box, "Pease's Principle," in Chapter 8.) I have never seen this hysteresis safety-factor technique outlined in print *for comparators*, so you can say you read it here first.

My suggestion for excess hysteresis is only a rule of thumb. Depending on your application, you might want to use even more hysteresis. For example, a comparator in an RC oscillator may operate with 1, 2, or 5 V of hysteresis, which means you can always use more than my "minimum amount" of excess hysteresis. Also, if you have a signal with a few millivolts of noise riding on top of it, the comparator that senses the signal will often want to have a hysteresis range that is two or three times greater than the worst-case noise.

Just the Right Touch

Comparators are literally very "touchy" components; that is, you can drastically alter their performance just by touching the circuit with your finger. And because comparators are so touchy, you should be prepared for the probability that your safety margin will change, for better or worse, when you go from a breadboard to a printed-circuit layout. There's no way you can predict how much hysteresis you'll need when your layout changes, so you just have to re-evaluate the system after you change it.

For faster comparators, such as the LM311, everything gets even touchier, and the layout is more critical. Yet, when several people accused the LM311 of being inherently oscillatory, I showed them that with a good layout, the LM311 is capable of amplifying any small signal, including its own input noise, without oscillating and without any requirement for positive feedback. One special precaution with the LM311 is tying the trim pins (5 and 6, normally) together to prevent AC feedback from the output (pin 7, normally), because the trim pins can act as auxiliary inputs. The LM311 data sheet in the National Semiconductor Linear Databook has carried a proper set of advice and cautions since 1980, and I recommend this advice for all comparators.

With comparators that are faster than an LM311, I find that depending on a perfect layout alone to prevent oscillation just isn't practical. For these comparators, you'll almost certainly need some hysteresis, and, if you are designing a sampled-data system, you should investigate the techniques of strobing or latching the comparator. Using these techniques can insure that there is no direct path from the output to the inputs that lasts for more than just a few nanoseconds. Therefore, oscillation may be evitable. Granted, heavy supply bypassing and a properly guarded PC-board layout, with walls to shield the output from the input, may help. But you'll probably still need some hysteresis.

For some specialized applications, you can gain advantages by adding AC-coupled hysteresis in addition to or instead of the normal DC-coupled hysteresis. (See Figure 9.3.) For example, in a zero-crossing detector, if you select the feedback capacitor properly, you can get zero effective hysteresis at the zero-crossover point while retaining some hysteresis at other points on the waveform. The trick is to let the capacitor's voltage decay to zero during one half-cycle of the waveform. But make sure that your comparator with AC-coupled hysteresis doesn't oscillate in an unacceptable way if the incoming signal stops.

Comparators *Do* Have Noise

Most data sheets don't talk about the noise of comparators (with the exception of the new NSC LM612 and LM615 data sheets), but comparators *do* have noise. Depending on which unit you use, you may find that each comparator has an individual "noise band." When a differential input signal enters this band slowly from either side, the output can get very noisy, sometimes rail-to-rail, because of amplified noise or oscillation. The oscillation can continue even if the input voltage goes back outside the range where the circuit started oscillating. Consequently, you could easily set up your own test in which your data for offset voltage, V_{OS}, doesn't agree with the manufacturer's measured or guaranteed values. Indeed, it can be tricky to design a test that *does* agree.

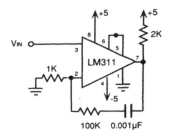

Figure 9.3. This zero-crossing detector has no DC hysteresis but 50 mV of AC-coupled hysteresis.

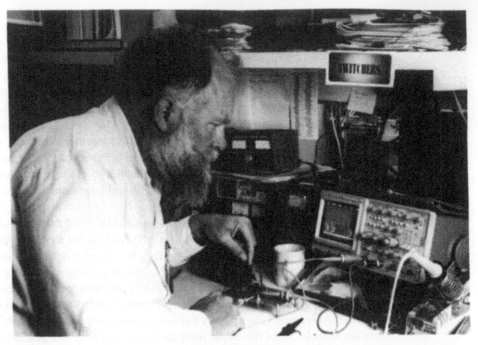

Figure 9.4. When oscillations get nasty, you need a scope with pinpoint triggering to help you study the problem.

For my tests of comparator V_{OS}, I usually set up a classic op-amp oscillator into which I build a specific amount of hysteresis and a definite amount of capacitance, so that the unit will oscillate at a moderate, controlled frequency. If you're curious, refer to Appendix D, which is not trivial.

Another way to avoid V_{OS} trouble with comparators is to use a monolithic dual transistor as a differential-amplifier preamplifier stage ahead of the comparator. This preamp can add gain and precision while decreasing the stray feedback from the output to the input signal. Refer to the example of a (fairly slow) precision comparator in LB-32 (Ref. 1).

Common-Mode Excursions Unpredictable

After curing oscillation, most complaints about comparators are related to their common-mode range. We at National Semiconductor's applications engineering department get many queries from engineers who want to know if they can violate comparators' common-mode specs. But they're not always happy with our answers. I guess the complaints are partly the fault of the manufacturers for not being clear enough in their data sheets.

By way of contrast, most engineers know well that an op amp's common-mode-voltage range, V_{CM}, is defined provided that both inputs are at the same level. This spec makes sense for an op amp because most operate with their inputs at the same level. But in most cases, a comparator's inputs are *not* at the same level. As long as you keep both inputs within the comparator's specified common-mode range, the comparator's output will be correct.

But if one input is within the common-mode range and the other is outside that range, one of three things could happen, depending on the voltages and the particular comparator involved. For some input ranges you can overdrive the inputs and still get perfectly valid response; for other input ranges, you can get screwy response but cause no harm to the comparator; and for others, you'll instantly destroy the comparator.

For example, for an LM339-type comparator running on a single 5-V supply, if *one* of its inputs is in the 0–3.5 V range, then the *other* input can range from 0–36 V without causing any false outputs or causing any harm to the comparator. In fact, at room temperature, the out-of-range input can go to –0.1 V and still produce the correct output.

But, heaven help you if you pull one of the inputs below the –0.1 V level, say to –0.5 or –0.7 V. In this case, if you limit the comparator's input current to less than 5 or 10 mA, you won't damage or destroy the comparator. But even if no damage occurs, the outputs of any or all of the comparators in the IC package could respond falsely. Current can flow almost anywhere within the IC's circuitry when the substrate diode (which is inherent in the device's input transistor) is forward biassed. It is this current that causes these false outputs.

We'll try to be more clear about V_{CM} specs in the future. Maybe next time at National Semiconductor, we'll phrase the spec sheet's cautions a little more vigorously. In fact—Ta-da—here is the correct phrase from the LM612 data sheet : "The guaranteed common-mode input voltage range for this comparator is $V^- \leq V_{CM} \leq (V^+–2.0 \text{ V})$, over the entire temperature range. This is the voltage range in which the comparisons must be made. If both inputs are within this range, of course the output will be at the correct state. If one input is within this range, and the other input is less than $(V^- + 32 \text{ V})$, even if this is greater than V^+, the output will be at the correct state. If, however, either or both inputs are driven below V^-, and either input current exceeds 10 μA, the output state is not guaranteed to be correct." And, this definition applies nominally to the LM339, LM393, and also to the LM324 and LM358 amplifiers if you are applying them as a comparator. So, you cannot say we are not trying to make our data sheets more clear and precise—even if it does sometimes take 20 years to get it just right.

Still, if you stay within their rated common-mode range, comparators are not that hard to put to work. Of course, some people disdain reading data sheets, so they get unhappy when we tell them that differential signals larger than ±5V will damage the inputs of some fast comparators. But this possibility has existed since the existence of the μA710, so you'll have to clamp, clip, or attenuate the input signal—differential or otherwise—so the fast comparators can survive.

An Unspoken Problem

Something else that does not usually get mentioned in a data sheet is common-mode slew problems. The good old LM311 is one part that is otherwise very well-behaved, but causes some confusion when common-mode slew problems arise. But to some extent all comparators can have these troubles. If one input suddenly slews up to exceed the other's level, you may see an unexpected, extra delay before the comparator's output changes state. This delay arises because the comparator's internal nodes do not slew fast enough for its outputs to respond. For example, a 10-V step can accrue an extra 100-ns delay compared with the delay for a 100-mV step. And if both inputs slew together, the output can spew out indeterminate glitches or false pulses even if the differential inputs don't cross over. Be careful if your circuit has comparator inputs of this sort, yet cannot tolerate such glitches.

Come to think of it, I get occasional complaints from engineers along the lines of, "I've been using this comparator for years without any trouble, but suddenly it doesn't work right. How come?" When we inquire, we find that the comparators have been operating very close to the edge of the "typical" common-mode range, well beyond with zero frequency. Although latched-up circuits demand troubleshooting, the good thing about them is that they sit right there and let you walk up to them and touch them. And you can measure every thing with a voltmeter to find out how they are latched. This state of affairs doesn't mean that troubleshooting them is easy, because sometimes you can't tell how the latched-up circuit got into its present state. And in an integrated circuit, there can be paths of carriers through the substrate that you can't "put your finger on."

The worst aspect of latched-up circuits is that some are destructive, so you can't just sit there and let them remain latched up forever. Two approaches for attacking destructive latches are:

- Turn off the power quickly, so the latched-up circuit cannot destroy anything. Try turning on power for short pulses and watching the circuit as it approaches the destructive latch condition. (See Chapter 2.)

- Use an adjustable current-limited supply with zero or small output capacitance, (such as the example in Chapter 2), so when the circuit starts to latch, the fault condition can easily pull the current-limited power supply's voltage down.

Another way to inadvertently generate a latched-up condition is to turn on the outputs of your multiple-output power supply in the "wrong" sequence. Some amplifiers and circuits get quite unhappy when one supply (usually the positive one) turns on first. Automatic power-supply sequencers can help you avoid this problem. An antireversal rectifier across each supply can help, too, and is always a good idea for preventing damage from inadvertently crossed-up power-supply leads or supply short circuits.

I used to get calls every few months from people who asked me if it was okay to ship (or launch) products that contained LM108s that may have had +15 V on their −15 V pins and vice versa. It was always painful for me to tell them, "Don't ship it— junk it. And, next time put antireversal diodes on each supply." Specifically, you should add these antireversal rectifiers across each bus in your system to protect the loads and circuits. Also add an antireversal rectifier across each power supply's output to protect the supplies (Figure 9.6). Some people think that leaving parts out is a good way to improve a circuit's reliability, but I've found that putting in the right parts in the right places works a lot better. Refer also to a running commentary and debate on this topic in Chapter 13.

If you have any doubt that your anti-oscillation fixes are working, try heating or cooling the suspected semiconductor device. In rare cases, passive components may be sufficiently temperature-sensitive to be at blame, so think about them, too. Even if a circuit doesn't get better when heated, it can get worse when cooled, so also take a peek at the circuit while applying some freeze mist.

My point is that merely stopping an oscillation is not enough. You must apply a tough stimulus to the circuit and see whether your circuit is close to oscillation, or safely removed from any tendency to oscillate. This stricture applies not only to regulators but also to all other devices that need oscillation-curing procedures.

For example, if a 47-Ω resistor in the base of a transistor cures an oscillation, but 24 Ω doesn't, and 33 Ω doesn't, and 39 Ω still doesn't, then 47 Ω is a lot more marginal than it seems. Maybe a 75-Ω resistor would be a better idea—just so long as 100- or 120- or 150-Ω resistors are still safe.

In other words, even though wild guesses and dumb luck can sometimes cure an oscillation, you cannot cure oscillations safely and surely without some thoughtful procedures. Furthermore, somebody who has an appreciation for the "old art" will probably have the best results.

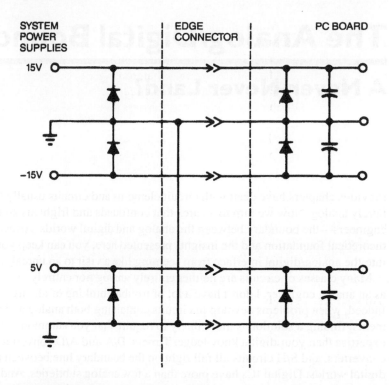

Figure 9.6. Installing antireversal diodes in the system power supplies *and also* on each PC board greatly reduces the chances of damaging the supplies or the circuitry with inadvertent short circuits or polarity reversals.

I certainly do not want to say that technicians can't troubleshoot oscillations simply because they don't know the theory of why circuits oscillate—that's not my point at all. I will only argue that a green or insensitive person, whether a technician or an engineer, can fail to appreciate when a circuit is getting much too close to the edge of its safety margins for comfort. Conversely, everyone knows the tale of the old-time unschooled technician who saves the project by spotting a clue that leads to a solution when all the brightest engineers can't guess what the problem is.

References

1. Linear Brief LB-32, Microvolt Comparator, in *NSC Linear Applications Book*, 1980–1990.

10. The Analog/Digital Boundary

A Never-Never Land?

Previous chapters have dealt with circuit elements and circuits usually thought of as purely analog. Now we turn to an area that confounds and frightens all too many engineers—the boundary between the analog and digital worlds. Armed with a solid theoretical foundation and the insights presented here, you can keep your journey into the analog/digital interface from seeming like a visit to an unreal world.

Many classes of circuits are neither entirely analog nor entirely digital. Of course, as an analog engineer, I don't have a lot of trouble thinking of all circuits as analog. Indeed, when problems develop in circuits containing both analog and digital elements, finding a solution is more likely to require that you summon your analog expertise than your digital knowledge. Timers, D/A and A/D converters, V/F and F/V converters, and S/H circuits all fall right on the boundary line between the analog and digital worlds. Digital ICs have more than a few analog subtleties. And even multiplexers, which you may have thought of as purely analog, have some quirks that result from their close association with the digital world.

Time for Timers

A timer is basically a special connection of a comparator and some logic, which is usually built with analog circuit techniques. The familiar 555 timer can do a lot of useful things, but it sure does get involved in a great deal of trouble. I'll treat the most classical fiascoes.

For one thing, people try to make timers with the crummiest, leakiest—usually electrolytic—capacitors. Then they complain because the timers are not accurate or their timing isn't repeatable. Some people insist on building timers to run for many seconds and then have trouble tweaking the time to be "exactly right." Sigh. These days I tell people, "Yes, you could make a 2-minute timer with an LM555 or a 10-minute timer with an LM322, but that would be *wrong*." Instead, you could make a simple 4-Hz oscillator using one-quarter of an LM324 or LM339 and cheap, small components. This oscillator can drive a CD4020 or CD4040; the last output of that counter, Q12 or Q14, can time very accurately and conveniently.

Such an arrangement is cheaper and much more accurate and compact than what you get if you blow a lot of money on a 47 μF polyester capacitor for a long-interval timer, or put up with the leakages of a tantalum capacitor, which no manufacturer wants to guarantee. In addition, in just a few seconds, you can trim the moderate-frequency oscillator by looking at an early stage of the divider; trimming a long-interval timer can take hours. The CMOS counters are inexpensive enough, and these days for 2- to 20-minute timer applications, I can usually convince customers not to buy the linear part. The LM555 data sheets tell you to avoid timing resistors with values higher than 20 MΩ. Nowadays, though, you can get a CMOS version (LMC555 or equivalent) or use a CMOS comparator or a CMOS op amp to work at

100 MΩ or more. Just be careful about board leakage and socket leakage—as you would with a high-impedance op-amp circuit. Then you can use a smaller, higher-quality capacitor.

Furthermore, it is a nontrivial statement that not all 555s work similarly; some manufacturers' 555s have different internal circuits and different logic flow charts. So be careful to check things out—555s from different manufacturers can act quite differently.

At high speeds, the timers don't just respond in a time 0.693 R × C; the response time is more like 0.693 R × (C + C_{STRAY}) + T_{DELAY}. Most books never mention this fact—most data sheets don't, either. So, although you can usually get a fast timer circuit to *function*, to get it to work the way you want it to, you still have to be careful. These designs are not always trivial, and Ref. 1 may help you avoid some pitfalls. A timer is, after all, just an aggregation of parts that includes a comparator, so many of the techniques you use with comparators work with timers, and vice versa.

Digital ICs: Not Purely Digital

Although timers are partly digital, the more classic digital ICs perform purely logical functions. Nevertheless, in the hands of a clever "linear" designer, some digital ICs can be very useful for performing analog functions. For example, CD4066 quad analog switches make excellent low-leakage switches and a 74C74 makes an excellent phase detector for a Phase-Locked Loop (Ref. 2). And not only is the price right—so is the power drain. Even when ordinary CMOS ICs aren't fast enough, you can often substitute a high-speed CMOS or 74ALS or 74AS counterpart to get more speed. I won't belabor the point; instead, I'll go straight to the litany of Troubles and Problems that you—whether an analog or a digital designer—can encounter with digital ICs.

First, unless proven otherwise, you should have one ceramic power-supply bypass capacitor in the range 0.02 to 0.2 μF—or even 1 μF, if the IC manufacturer requires it—for each digital IC, plus a tantalum capacitor in the range 2 to 10 μF for every two, three, or four ICs. The ceramic capacitors provide good local high-frequency bypassing; the tantalum parts damp out the ringing on the power-supply bus. If you can't use a tantalum capacitor, you can use 10 or 20 μF of aluminum electrolytic, *or* if you are desperate you can try a 1 or 2 μF extended-foil Mylar unit in series with a 1 Ω carbon resistor, to provide the needed lossiness. If your linear circuit really depends on clean, crisp digital outputs (CMOS outputs make dandy square-wave generators, as long as the power supply isn't ringing and bouncing) you may even want more bypassing—possibly hundreds of microfarads.

Floating Inputs Can Leave You at Sea

On TTL parts, you can leave an unused input floating and it will normally go HIGH; on CMOS, you *must* tie unused inputs (such as the preset and clear inputs of a flip-flop) to the positive supply or ground, as appropriate. Otherwise, these inputs will float around and give you the screwiest intermittent problems. Also, when these inputs float, for example, on unused gates, they can cause considerable unwanted power drain and self-heating.

With CMOS, people used to tell you that you can use an inverter as an amplifier by tying a few megohms from the input to the output. At low voltages, you can make a mediocre amplifier this way, but when the supply voltage is above 6 V, the power drain gets pretty heavy and the gain is low. I don't recommend this approach for modern designs.

Many years ago, people used to tie the outputs of DTL or open-collector TTL gates together to form a "wired OR" gate. This practice has fallen into disrepute as it supposedly leads to problems with troubleshooting. I don't know what other reason there is for not doing it, except to avoid acting like a nerd. However, an open-collector output with a resistive pull-up *is* slower than a conventional gate *and* wastes more power.

A couple engineers chided me, that if you let TTL or DTL inputs float, that may appear to work OK for a while, but when you get all the signal busses packed in together, the unused inputs may be driven to give a false response—not consistently but intermittently. So, it is bad practice to let your TTL inputs float. It is also not quite correct to tie those inputs to the +5-V bus. Tie them up toward +5 V through 1 k. Then a momentary +7-V transient on the supply bus may do less harm, less damage.

When digital-circuit engineers have to drive a bus for a long distance, say 20 or 30 inches, they use special layouts, so the bus will act like a 75 Ω or 93 Ω stripline. They also add termination resistors at one or both ends of the bus to provide damping and to cut down on reflections and ringing. When *you* have to drive long lines in an analog system, *you* must do the same. Note that for really fast signals, digital designers don't even lay out their PC traces with square corners; they bend the foil around the corner in a couple of 45° turns. Many digital engineers are not just bit-pushers; they've been learning how to handle real signals in the real world. They are actually pretty expert in some analog techniques, and analog engineers can learn from them.

Perfect Waveforms Don't Exist

Even though many digital engineers are familiar with real problems, they often sketch the waveforms from gates and flip-flops showing nice, crisp, vertical rises and showing the output of a gate changing at the same time as the input. But smart engineers are aware that when it comes down to the fine print, they must be prepared to admit that these waveforms have finite rise-times and delays. These nit-picking details are very important when your signals are in a hurry.

For example, if the data input of a D flip-flop rises just *before* you apply the clock pulse, the output goes high. If the data input rises just *after* you apply the clock pulse, the output goes low. But if the D input moves at *just* the wrong time, the output can show "metastability"—it can hang momentarily halfway between HIGH and LOW and take several dozen nanoseconds to finally decide which way to go. Or, if the data comes just a little earlier or later, you might get an abnormally narrow output pulse— a "runt pulse."

When you feed a runt pulse to another flip-flop or counter, the counter can easily respond falsely and count to a new state that might be illegal. Thus, you should avoid runt pulses and make sure that you don't clock flip-flops at random times. Fig 10.1a contains an example of a D flip-flop application that can exhibit this problem. When the comparator state changes at random times, it will occasionally change at precisely the wrong time—on the clock's rising edge—making the output pulse narrower or wider than normal. In certain types of A/D converters, this effect can cause nonlinearity or distortion. A good solution is to use a delayed clock to transfer the data into a second flip-flop, as in Fig 10.1b.

A glitch is an alternate name for a runt pulse. A classic example of a glitch occurs when a ripple counter, such as a 7493, feeds into a decoder, such as a 7442. When the

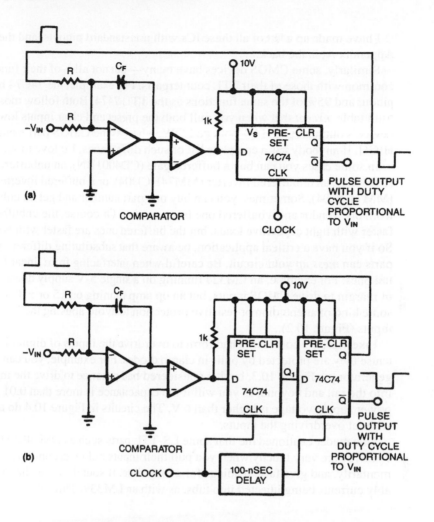

Figure 10.1. Runt pulses cause problems in this simple ADC (a). The comparator state changes at random times. Occasionally, the state will change at precisely the wrong time—on the clock's rising edge—making the output pulse narrower or wider than normal. You can solve the problem by using two flip-flops with the clocks separated by a delay.

counter makes a carry from 0111 to 1000, for a few nanoseconds the output code will be 0000, and the decoder can spit out a narrow pulse of perhaps 6–8 ns in duration corresponding to 0000. Even if you are observing with a good scope, such a pulse can be just narrow enough to escape detection. If the decoder were merely feeding an LED display, you would never see the sub-microsecond light pulse, but if the decoded output goes to a digital counter, a false count can occur. In digital systems, engineers often use logic analyzers, storage scopes, and scopes with very broad bandwidths to look for glitches or runt pulses and the conditions that cause them. In analog systems, you may not have a logic analyzer, but these nasty narrow pulses often do exist, and you have to think about them and be prepared to look for them.

Another thing to know about digital ICs is that *many* CMOS ICs have the same pinouts as TTL parts. For example, the 74193, 74LS193, and 74C193 have the same pinouts. On the other hand, some of the older CMOS parts have pinouts that differ from those of similarly numbered TTL devices. The 74C86's pinout is the same as the 74L86's but differs from the 7486's. Beware!

I have made up a list of all these ICs with nonstandard pinouts, and they are in Appendix A, in the back....

Similarly, some CMOS devices have many—but not all—of their functions in common with those of their TTL counterparts. For example, the 74C74 has the same pinout and 95% of the same functions as the TTL 7474. Both follow mostly the same truth table, *except* that when you pull both the preset and clear inputs low, the TTL device's outputs (Q and $\overline{Q}$) both go LOW, whereas the CMOS part's outputs both go HIGH. If anybody has a complete list of such differences, I'd love to see a copy.

In some cases you can buy a buffered gate (CD4001BN), an unbuffered gate (CD4001), an unbuffered inverter (MM74HCU04), or a buffered inverter (MM74HC04). Sometimes, you can buy one part number and get an unbuffered part from one vendor and a buffered one from another. Of course, the unbuffered parts are faster with light capacitive loads, but the buffered ones are faster with heavy loads. So if you have a critical application, be aware that substituting different vendors' parts can *mess up* your circuit. Be careful when interfacing from linear ICs into digital ones. For example, an LM324 running on a single 5-V supply doesn't have a lot of margin to drive CMOS inputs, but an op amp running on ±5 or ±10 V would need some kind of attenuation or resistive protection to avoid abusing the logic-device inputs (Figure 10.2).

Likewise, it's considered bad form to overdrive the inputs of digital ICs just because they are protected by built-in clamp diodes. For example, you can make a pulse generator per Figure 10.3, but it's considered bad practice to drive the inputs hard into the rail and beyond, as you will if the capacitance is more than 0.01 μF or the power supply voltage is higher than 6 V. The circuits in Figure 10.4 do as good a job without overdriving the inputs.

One reader cautioned me that some LS-TTL parts such as DM74LS86 and 74LS75s are very touchy when you pull their inputs below ground even micro-momentarily, and give false readings for a long time. It sounds to me that there are probably currents being injected into tubs, as with an LM339: Thus, these are unhappy

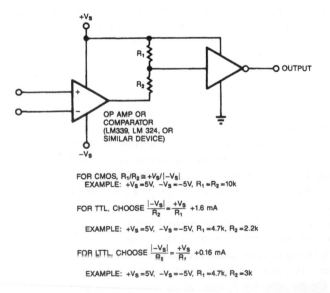

FOR CMOS, $R_1/R_2 \cong +V_S/|-V_S|$
    EXAMPLE: $+V_S = 5V$, $-V_S = -5V$, $R_1 = R_2 = 10k$

FOR TTL, CHOOSE $\frac{|-V_S|}{R_2} = \frac{+V_S}{R_1} + 1.6$ mA

    EXAMPLE: $+V_S = 5V$, $-V_S = -5V$, $R_1 = 4.7k$, $R_2 = 2.2k$

FOR LTTL, CHOOSE $\frac{|-V_S|}{R_2} = \frac{+V_S}{R_1} + 0.16$ mA

    EXAMPLE: $+V_S = 5V$, $-V_S = -5V$, $R_1 = 4.7k$, $R_2 = 3k$

Figure 10.2. Driving logic from an op amp operating from the usual large supply voltages requires an attenuator between the amplifier and the logic IC. The equations show how to calculate the attenuator ratios.

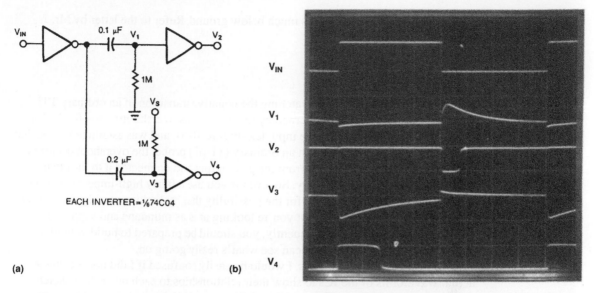

Figure 10.3. This CMOS pulse generator (a) is not recommended because, with the values shown, it overdrives the gate inputs excessively—as the waveforms of (b) indicate.

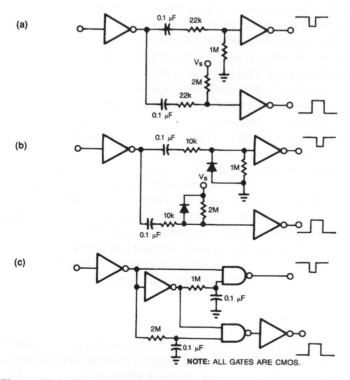

Figure 10.4. The addition of resistors to the circuit of Figure 10.3 (a) helps reduce overdrive, but the addition of diode clamps in the shunt leg of the attenuators (b) is even more effective. If you have two 2-input NAND gates available, the circuit of (c) is the best implementation.

parts if you overdrive the inputs much below ground. Refer to the letter by Mr. J. Koontz in Chapter 13.

A Time to Ask Probing Questions

A number of years ago, I was watching the negative transition of an ordinary TTL gate, and I was especially concerned by the way it was overshooting to –0.4 V. I set up an attenuator with 1 pF in the input leg (Figure 10.5), and was astounded to see that if I looked at the waveform with an ordinary (11-pF) probe, the overshoot occurred, but if I disconnected the probe from the gate output and connected it to the attenuator output, the overshoot went away. So, even if you use a fairly high-impedance probe, you should always be prepared for the possibility that by looking at a signal, you can seriously affect it—even if what you're looking at is as mundane and supposedly robust as a TTL output. Consequently, you should be prepared to build your own special-purpose probes, so you can see what's really going on.

When I work with digital ICs, I would be easily confused if I did not sketch the actual waveforms of the ICs to show their relationships to each other. So I sketch these waveforms on large sheets of quadrille paper (1/4-in. grid) to produce something I call a "choreography" because it maps out what I want all the signals to do and exactly where and when I require them to dance or pirouette.... When the system gets big and scary, I sometimes tape together two or three or four sheets horizontally and as many sheets as I need vertically. Needless to say, I am not very popular when I drag one of these monsters up to the copying machine and try to figure out how to make a copy. Figure 10.3b is a small example.

NOTE, when I first published this Figure 10.3b in EDN magazine in 1989, the sketch was printed with an error; some of the pulses were positioned at the wrong time. And did EDN make the error? Not at all! *I* drew it wrong and the error wasn't caught until after publication when a kid engineer suggested it might be erroneous. He was right. How embarrassing. It would have been even worse if a whole lot of people had called to correct me. That just goes to show, if you stand on a big soapbox and rant and holler, people will often think you know what you are talking about. They stop looking for mistakes.... and that's a mistake. Bigwigs make mistakes—and wanna-be big-wigs, too. Embarrassing....

Maybe the guys who design really big digital ICs can get along without this choreography technique; maybe they have other mnemonic tools, but this one works for me. I first developed this approach the time I designed a 12-bit monolithic ADC, the industry's first, back in 1975. I had this big choreography, about 33 inches square, and the circuit worked the first time because the choreography helped me avoid goofing up any digital signals. Right now I'm working on a system with one choreography in nanoseconds and tenths of nanoseconds linked to a second one scaled in microseconds and a third one scaled in seconds. I hope I don't get lost.

Of course, this tool is partly for design, but it's also a tool for troubleshooting—and for planning, so you can avoid trouble in the first place.

D/A Converters Are Generally Docile

D/A converters are pretty simple machines, and they can usually give excellent results with few problems. If the manufacturer designed it correctly and you are not misapplying it, a DAC usually won't cause you much grief.

One area where DACs can cause trouble, however, is with noise. Most DACs are not characterized or guaranteed to reject high-frequency noise and *jumps* on the

Figure 10.5. An ordinary high-impedance probe can cause TTL outputs to appear to overshoot when you look at them, but *not* when you are not looking at them. You can eliminate this effect by making your own very-high-impedance probe, that presents only a 1-pF capacitive load.

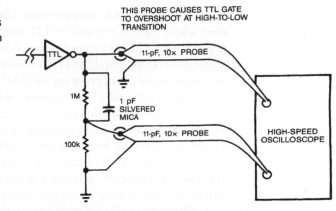

Figure 10.6. The author irritates co-workers when he carries one of his large "choreographies" to the photocopier and tries to figure out how to duplicate such a large drawing. In fact, some say that he irritates co-workers most of the time.

supply voltages. In some cases, the *DC* rejection can be 80 or 100 dB, but high-frequency noise on a supply can come through to the output virtually unattenuated. So you must plan your system carefully. It might be a good idea, in a critical application, to use a completely separate power-supply regulator for your precision DAC. At least you should add plenty of good power-supply bypass capacitors right at the power-supply pins—ceramic and tantalum capacitors.

Sometimes when you feed signals to a DAC without passing them through buffers, the noise, ringing, and slow settling of the digital signals can get through to the analog side and show up on the DAC output. Nobody has a spec for rejection of the noise on DAC bit lines in either the HIGH or LOW state. Maybe vendors should specify this parameter, because some DACs are good and some aren't. I even recall a case where I had to preload the TTL outputs of a modular DAC's internal storage register with a 2 kΩ resistor from each line to ground. Otherwise they would over-shoot when going HIGH and then recover with a long slow tail, an attenuated version of which would then appear on the DAC output.

On-chip buffers at a DAC's input can help cut down feedthrough from the bit lines to the analog output, but will not completely eliminate it. The bus can move around incessantly, and capacitive coupling or even PC-board leakage will sometimes cause significant crosstalk into the analog world. Even IC sockets can contribute to this noise. If you could prove that such noise wouldn't bother your circuit, you could forget about it. The problem is that you can only make meaningful measurements of such effects on an operating prototype—computer modeling isn't going to simulate *this*.

Multiplying DACs are popular and quite versatile. However, a multiplying DAC's linearity can be degraded if the output amplifier's offset voltage isn't *very* close to zero. I've heard this degradation of linearity estimated at 0.01% per millivolt of offset. Fortunately, low-offset op amps are pretty cheap these days. At least a low-offset op amp is cheaper than a trim-pot.

Another imperfection of any multiplying DAC is its AC response for different codes. If you put in a 30-kHz sine wave as the reference, you shouldn't really be surprised if the gain from the reference to the output changes by more than 1 LSB when you go from a code of 1000 0000 to a code of 0111 1111. In fact, if the frequency is above 5 kHz, you may find a 0.2% or larger error because the multiplying DAC's ladders, whose attenuation is a linear function of the input code at DC, become slightly nonlinear at high frequencies due to stray capacitance. The non-linearity can be 0.2%, and the phase change as you vary the input code can exceed 2°, even with a 5-kHz reference. So don't let these AC errors in multiplying DACs surprise you.

Another problem with DACs is the output glitch they can produce when going from one code to an adjacent one. For example, if a DAC's input code goes from 1000 0000 to 0111 1111 and the delay for the rising bits is much different from that for the falling bits, the DAC output will momentarily try to go to positive or negative full scale before it goes to a value corresponding to the correct code. Though well-known, this problem is a specialized one. One possible solution calls for precisely synchronous timing. Multiple fast storage registers can also help to save the day. But if the best synchronous timing is not good enough, a deglitcher may be the solution.

ADCs Can Be Tough and Temperamental

Like DACs, many A/D converters (ADCs) do exactly what they are supposed to, so what can go wrong? Most problems involve a characteristic that is mentioned on too few data sheets: Noise. When an analog signal moves slowly from one level to an-other, it would be nice if the ADC put out *only* the code for the first voltage and then, at the appropriate threshold, began to produce *only* the code for the other voltage. In practice, there is a gray area where noise causes codes to come up when they shouldn't. On a good ADC, the noise can often be as low as 0.1 or 0.05 LSB p-p. But when you come to a worst-case condition (which with successive-approximation converters often occurs at or near a major carry   for example, where the output changes from 1000 0000 to 0111 1111), the noise often gets worse, sometimes climbing to 0.5 LSB p-p or more. I wouldn't want to buy an ADC without knowing how quiet it was. I'd have to measure the noise myself, as shown in Figure 10.7, because virtually nobody specifies it. That's not to say all ADCs are bad, just that manufacturers don't make much noise about noise. Ron Knapp of Maxim wrote a nice explanation of this ADC noise measurement technique in EDN recently (Ref. 3). I recommend his article on this subject.

Most ADC data sheets spell out that the only correct way to test or use an ADC is with the analog signal's ground, the digital supply's ground, and the analog supply's ground tied together right at the ground pin of the ADC. If you don't or can't interconnect the grounds at the specified point, all bets are off.

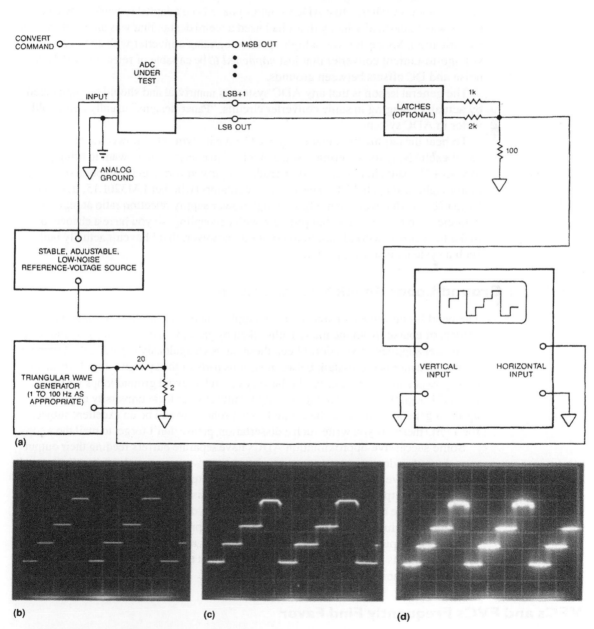

Figure 10.7. A reference source, a triangular-wave generator, and a scope are the major building blocks of an ADC crossplot tester (a) that can reveal how much noise a converter adds to the signal it is digitizing. In (b) the noise performance is ideal, whereas in (c) it is merely acceptable. In (d) the noise performance is unacceptable.

With ADCs, Paper Designs Aren't Adequate

On one 10-bit ADC I designed, when the customer found some problems that I couldn't duplicate in my lab, I bought one plane ticket for me and one for my best scope. After a few hours we arrived at the scene, and in less than an hour I had the problem defined: The customer expected our converter to meet all specs with as much as 0.2 V DC plus 0.2 V AC, at frequencies as high as 5 MHz, between the analog ground and the digital ground!! Outrageous!! Amazingly, our architecture was such that by deleting one resistor and adding one capacitor, I could comply with the customer's wishes. Most ADCs couldn't have been adapted to work—the customer was fantastically lucky that I had used a weird design that was amenable to this modification. My design was a high-speed integrating converter with an input voltage-to-current converter that just *happened* to be capable of rejecting wideband noise and DC offsets between grounds.

The general lesson is that any ADC system is nontrivial and should be engineered by actually plugging in some converter circuits. "Paper designs" usually don't hold water in ADC systems.

To beat the requirement that every ADC have its own set of power supplies dedicated exclusively to powering just the single converter, you may want to bring power to your PC boards in unregulated or crudely regulated form, and then put a small regulator right near each ADC. These small regulators (whether LM320L15, μA78L05, LM317L, or whatever) do not have a high power-supply rejection ratio at high frequencies. You can resolve that problem with decoupling, so you have a chance to make the scheme work. I hasten to point out, however, that I haven't actually built such a system myself very often.

Don't Let Ground Loops Knock You for a Loop

The need for multiple (separate) power supplies, or at least multiple regulators, comes, of course, from the many paths taken by ground currents flowing to and from the power supplies. If you don't keep these paths scrupulously separate, the ground loops can cause bad crosstalk between various parts of the system—low-level analog, high-power analog, and digital. So be very careful to avoid ground loops when you can. Although the electrical engineering faculty at your local university might not agree, a general solution to the ground-loop problem would be an excellent subject for a PhD thesis. If you write such a dissertation, please don't forget to mail me a copy.

Some successive-approximation ADCs have separate buffers feeding their output pins, but other designs try to save money, parts, power, or space by using the internal registers to drive both the internal DAC and the output pins. In this case, external loads on the outputs can cause poor settling and noise and can thus degrade the performance of the converter. If you're using ADCs, you should find out if the outputs are connected directly to the DAC. Sometimes, as previously mentioned, a preload on each bit output can help to accelerate settling of an ADC's internal DAC. After all, TTL outputs must be able to drive more current than their DC specs state—they have to meet their AC specs.

VFCs and FVCs Frequently Find Favor

The voltage-to-frequency converter (VFC) is a popular form of ADC, especially when you need isolation between the analog input and digital outputs. You can easily feed a VFC's output pulse train through an optocoupler to achieve isolation between different ground systems. The VFC can cover a wide range with 14 to 18 bits of dynamic range. The less expensive VFCs are slower; the faster ones can be expensive. Most VFCs have excellent linearity, but the linearity depends on the timing capacitor having low dielectric absorption. Teflon makes the best VFC timing capacitors, but polystyrene, polypropylene, and ceramic capacitors with a C0G characteristic are close behind. (Refer to the LM131/LM331 data sheet, for examples and notes.)

Trimming a VFC to get a low temperature coefficient is not easy because the overall temperature coefficient depends on several components, including the reference, as well as various timing delays. See Ref. 4 for VFC trimming procedures or, at least, to appreciate how much effort is involved when you buy a well-trimmed unit.

The Other Way Back

Frequency-to-voltage converters (FVCs) are often used as tachometers or in conjunction with a VFC and an optoisolator to provide voltage isolation in an analog system. FVCs are about as linear as VFCs and about as drifty, so the temperature trimming problem is the same as for a VFC. One exception is if you're using cascaded VFC/FVC pairs in which both circuits are in the same location and at the same temperature. In that case, you can often get by with trimming only one of the pair, or by just making sure the TCs match!

Another problem with FVCs is that you often want the response to be as fast as possible but need to keep the ripple low. The design of a filter to accomplish both objectives will, of course, be a compromise. My rule of thumb is that you can keep the ripple down to about 0.01% of the $V_{fullscale}$, but with the simplest filters, you must keep the carrier at least 100 times the F_{min}. With more sophisticated filtering, such as two Sallen-Key filters cascaded, the −3-dB point can be 1/10 of the slowest carrier. For example, with a carrier frequency in the range 5–10 kHz, the signal can go from DC to 500 Hz (Ref. 5). If you need still faster response, see Ref. 2, which shows in a cookbook circuit how to use a simple phase-locked loop to make a surprisingly quick FVC.

S/H Circuits: Electronic Stroboscopes

A VFC produces an output proportional to the average value of its analog input during the conversion. If you need to digitize rapidly changing signals, for example, to reconstruct waveforms in the digital domain, you need a different type of ADC and you almost always have to precede it with a sample-and-hold circuit. Designing S/H circuits is a complicated, challenging endeavor. Meeting exacting specs often requires an expensive module or hybrid circuit. A major problem of S/H circuits is dielectric absorption, or "soakage," in the hold capacitor (Ref. 6). If you need to run a relatively short sample time with a long hold time and if the new output voltage can vary considerably from the previous sample, the soakage may be your biggest problem. For example, if an S/H circuit acquires a new voltage for 5 μs and then holds it for 500 μs, you can tell approximately what the previously held signal was because the new V_{out} can shift by 2–3 mV—the amount and direction depend only on the value of the previous signal. And that's for an expensive Teflon hold capacitor—most other capacitors have soakages three to five times worse. If the timing, frequency, and rep-rate don't change, you may be able to add a circuit to provide some compensation for the soakage (Ref. 7); but the problem isn't trivial, and neither is the solution. Cascading two S/H circuits—a fast one and a slow one with a big hold capacitor—won't help the soakage but will tend to minimize the problem of leakages.

Some people wish that a S/H circuit would go from sample to hold with a negligible jump, or "glitch." Although you can build such a circuit, it's a lot more difficult than building a more conventional S/H circuit. You usually find glitch-free S/H circuits only in "deglitchers," which are more expensive than most S/H circuits. Several module and hybrid manufacturers provide this kind of precision device. Even though it doesn't settle out instantly, a deglitcher is fast and consistent in its settling. However, it still does take some time to settle within 5 mV.

Aperture Time Still Causes Confusion

There's one area of specsmanship where the S/H circuit is clouded in confusion. That area is the aperture-delay specification. (Maybe someday I'll write a data sheet and drive away the cloud. Ask me for a data sheet on the LF6197....) One technique for measuring and defining aperture delay is to maintain V_{in} at a constant level and issue the HOLD command. If after a short delay, V_{in} jumps by a few volts, the smallest spacing between the HOLD command and the V_{in} jump that causes no false movement of V_{out} is one possible definition of the $t_{APERTURE\ DELAY}$.

Another way of defining and measuring aperture delay is to let V_{in} move smoothly at a well-defined rate. Shortly after you issue the command to switch the circuit to the HOLD mode, V_{out} stops changing. The value at which V_{out} stops corresponds to the value of V_{in} at a particular point in time. You can define the aperture delay as the difference between this point and the time at which the mode-control signal crossed the logic threshold. The uncertainty in the value of the aperture delay is then the aperture uncertainty. Depending on how the circuit was optimized, that delay can be positive or negative or practically zero—perhaps only 1 ns or less. Now, will the real definition of aperture time please stand up?

I think that both of the characteristics I have described are of interest to people at different times. But, how can you avoid the problem of a person expecting one of these characteristics and actually getting the other? I invite your comments on who wants to buy which characteristic, and where to find a definition. I've looked in military specs and at many data sheets, and the issue still seems pretty unclear.

Another instance in which a S/H circuit can have trouble is when its output is connected to a multiplexer, for example, when multiple S/H circuits drive a single ADC to achieve simultaneous sampling of many channels of dynamic analog data. If the multiplexer, which had been at a voltage of, say, +10 V, suddenly connects to the output of a S/H circuit whose output is at −10 V, the circuit's output will twitch and then may jump to a false level because the multiplexer will couple a little charge through the S/H circuit into the hold capacitor. The industry-standard LF398 is fairly good at driving multiplexers, but if you get a big enough capacitance on the multiplexer output—perhaps 75 pF—and it's charged to a voltage more than 10 V away from the S/H circuit's output voltage, even the LF398's output can jump. I don't have a real solution for this problem, but if you are aware that it can happen, at least you won't tear out all your hair trying to guess the cause. You will recognize the problem, and *then* tear out your hair. About all you can do is try to minimize the capacitance on the output of the multiplexer. One way to do this is by using a hierarchical connection of submultiplexers.

Not Much Agreement on Acquisition-Time Definition

Another area of S/H-circuit confusion is acquisition time. I have seen at least one data sheet that defined acquisition time as the time required to go from HOLD to SAMPLE and for the output to then settle to a value corresponding to a new value of

V_{in} *in the SAMPLE mode.* However, the outputs of many S/H circuits can settle to a new DC value faster than the hold capacitor charges to the correct value. If the S/H doesn't have to go back to HOLD, the data may give false results, even if the output seemed to give the correct answer when it was still in SAMPLE mode. To avoid confusion, *we* define acquisition time as the pulse width required for precise sample-and-hold action. If the circuit SAMPLEs and settles and then goes into HOLD and gives you the wrong answer, the SAMPLE pulse should have been wider—right? Right.

There may be some S/H circuits whose output voltage won't change if you switch them to the HOLD mode as soon as their output reaches a value that corresponds to a new V_{in}. But if I had an analog switch that couldn't hold at all, I could still get it to "acquire" a signal according to the data-sheet definition just cited. I consider the test implied by that definition to be too easy. I believe some users and manufacturers in this field agree with my definition, but the situation isn't really clear. (I would appreciate reader comments. You folks are getting all sorts of good ideas from me, and if you have some good comments, it's only fair that you bounce them off me.)

E Pluribus Unum: The Multiplexer

Another type of circuit that depends on analog switches is the analog multiplexer. As mentioned already, a multiplexer can draw big transients if you suddenly connect it across big signals at low impedances. So be careful not to overdo operating a multiplexer in this manner, as excessive current could flow and cause damage or confusion. It's well known that multiplexers, like most other forms of analog switches, are imperfect due to leakages, on-resistance, and response time. But they are popular and won't give you much trouble until you turn the power supplies OFF and keep the signals going. I recall that in the past few years, at least one or two manufacturers have brought out new designs that could survive some fairly tough over-voltages with the power removed. I'm not sure what the designs involved other than adding thin-film resistors and diode clamps on the inputs—ahead of the FET switches. But if *you* add discrete resistors ahead of any monolithic multiplexer's inputs, the resistors can help the multiplexer survive the loss of power.

One other problem with multiplexers is that you don't have a whole lot of control over the break-before-make margin. And if you should want make-before-break action, I don't think it's an available option. So, sometimes you may have to "roll your own" multiplexer.

If your signal levels are less than 15 V p-p, you may be able to use the popular CD4051 and CD4053 multiplexers and the CD4066 CMOS analog switches, which are inexpensive and quick and usually exhibit low leakage. However, if you need a guarantee of very low leakage, you will have to test and select the devices yourself, as many people do.

Digital Computers

To avoid making unpleasant comments, I will simply say that I hope somebody else writes a good book on troubleshooting these.

Software

NO comment....

So, we take leave of the analog/digital world—sort of. In the next chapter, we'll visit another area of great importance to analog/digital electronics, but it is a purely linear region, perhaps the most purely linear: References. Armed with knowledge about references, we'll move on to the troubleshooting of power electronics, including switching regulators.

References

1. Jung, Walter, *555 Timer Cookbook*, Howard Sams and Co, Indianapolis, IN, 1977.

2. Pease, Robert A., "Wideband phase-locked loops take on F/V-conversion chores," *EDN*, May 20, 1979, p. 145. (Also available as AN-210 in NSC's Linear Applications Book, 1986, 1989, etc. "New Phase-locked-loops Have Advantages as Frequency-to-Voltage Converters (and more).")

3. Knapp, Ron, "Evaluate your ADC by using the crossplot technique," *EDN*, November 10, 1988, p. 251.

4. Pease, R. A., "Versatile monolithic V/Fs can compute as well as convert with high accuracy," *Electronic Design*, December 6, 1978, p. 70. (Also available as Appendix D in National Semiconductor Corp. *Linear Applications Handbook*, Santa Clara, CA, 1986, p. 1213.)

5. Pease, R. A., "V/F-converter ICs handle frequency-to-voltage needs," *EDN*, March 20, 1979, p. 109. (Also available as Appendix C in National Semiconductor Corp. *Linear Applications Handbook*, Santa Clara, CA, 1986, p. 1207.)

6. Pease, R. A., "Understand capacitor soakage to optimize analog systems," *EDN*, October 13, 1982, p. 125.

7. National Semiconductor Corp., *Linear Databook 2*, Santa Clara, CA, 1986, p. 5-5.

11. Dealing with References and Regulators

As the self-appointed Czar of Band Gaps, I am impelled to continue this book with words of wisdom on voltage references, regulators, and start-up circuits. I also provide warnings against assumptions about worst-case conditions.

Voltage references and regulators have internal features that make them relatively immune to problems. But, as with other designs, if you ignore the details, you'll be headed for Trouble. Some designs incorporating these parts, such as switching power supplies, are not for the novice.

Many voltage references are based on band-gap circuitry, but some of the best references are based on buried zener diodes. If your power supply's output is in the 8–12 V range, or higher, zener-diode references such as LM329, LM399, or LM369 can provide high stability, low noise, and a low temperature coefficient. If your power supply is a lower voltage (in the range from 8 down to 1.1 V) you can find band-gap references that put out a stable voltage anywhere from 0.2 to 5 V with creditable efficiency and economy. These band-gap references feature as low a temperature coefficient as you'd probably ever be willing to pay for—as good as 20 or 10 ppm /°C. (They also feature enough noise so that a little filtering can make a big improvement.)

A good buried-zener-diode voltage reference is inherently more stable over the long term than is a band-gap one—good zener designs change only 5 to 10 ppm per month. However, if you want the best stability possible, it's only fair to age, stabilize, and burn-in the references first. Also, you must screen out the ones that just keep "walking" away from their initial values by 10 to 20 ppm every week—there are always a few "sports" that are driftier than the rest. Unfortunately, there's no quick and easy test to distinguish between the drifty ones and the stable ones, except for taking measurements for many hundreds of hours.

Regulators Are *Almost* Foolproof

In the last 10 years, IC voltage regulators have gotten pretty user-friendly. Many people use them with no problems at all. Still, my colleagues and I get at least one call every month about a regulator working poorly. The indignant caller complains, "It's getting hot." We ask, "How big is your heat sink?" The indignant voice responds, "What do you mean, heat sink?" I credit all of you readers with enough smarts to recognize that you can't put a whole lot of power into a little regulator unless you secure it to a sufficient heat sink or heat fins. Then, there really aren't too many things that are likely to go wrong, because voltage regulators have just about every feature for protection against the world's assaults.

You'll have problems with regulators when you don't provide the required, specified output bypassing. Most negative regulators and some other types, such as low-dropout regulators, require an electrolytic bypass capacitor from the output to ground. If you insert a tantalum capacitor, you may be able to get away with a value

Figure 11.1. When you're the Czar of Band Gaps, people look up to you.

of 1 or 2 μF; if you use an aluminum electrolytic capacitor, you can get away with 20 to 100 μF, or whatever the data sheet spells out. But in all cases, on all the parts I know, an electrolytic capacitor *will* work, and a film or ceramic capacitor *won't* work—its series resistance is just too small. Now, if you put a 1 Ω resistor in series with a 1 μF ceramic capacitor, the filtering will probably be adequate around room temperature; the loss factor is then similar to a tantalum capacitor. But if you take it to –40 or +100 °C, the ceramic capacitor's value will shrink badly (refer to Chapter 4 on capacitors) and the regulator will be unhappy again. It may start oscillating, or it might just start ringing *really badly*.

Recently one of our senior technicians was helping a customer with some applications advice. He found that the AC output impedance of an LM317 was changing considerably as a function of the load current coming out of the output transistor. We had always assumed that the curve in the LM117 data sheet was invariant versus current load—that was a mistake. Then we found that every other monolithic regulator has the same sliding scale of output inductance. For additional notes on this phenomenon, I recommend the Erroll Dietz article (Ref. 1) which I have included as Appendix C because this tendency of the output inductance to be modulated by the output current may help to explain why regulators are happy in some cases but grouchy in other similar situations.

Another regulator problem can occur when you add an external transistor to increase the output current. Since this transistor adds gain at DC, it's not surprising that you have to add a big filter capacitor on the regulator's output to prevent oscillation. Some of the applications in old National Semiconductor data sheets recommend specific values for the filter capacitor, and specific types for the boost transistors, but some of these circuits are quite old. When customers find out that 2N3234s are no longer available, they're likely to substitute a more modern transistor that has a faster response and is likely to oscillate. In this case, a customer might complain about the DC output's "bad load regulation" as the regulator is forced into and out of oscillation. (Whoever said you don't need an oscilloscope to check out DC problems?)

When these customers ask for help, I not only explain how to stop the oscillation, but I give them Pease's Principle (see the box, "Pease's Principle" in Chapter 8). However, these days most engineers find it's better to use a *bigger* regulator (LM350 at 3 A, LM338 at 5 A) because if you just add on an external transistor, you cannot protect it from overheating. Consequently the external power transistor has lost favor.

Too Much Voltage Leads to Regulator Death

You *can* kill any regulator with excessive voltage. So if you're driving inductive loads, or if your circuit has an inductive source, make sure to have a place for the current to go when the normal load path changes. For example, if you're using the LM350 as a simple battery charger with only a few microfarads of filter capacitor on the input, a short between the output and ground is usually disastrous: When the regulator tries to draw an increasing amount of current from the transformer and then goes into current limit, the inductance of the transformer will give you marvelous 80-V transients, which then destroy the LM350. The solution is to put 1000 μF—rather than just 1 or 10 μF—across the input.

Users get accustomed to seeing regulators with output noise of about 0.01% of the rated DC output. They get indignant when the noise doubles or triples due to 1/f or popcorn noise. The chances of finding a noisy regulator are quite small, so when some noisy ones *do* show up, it's a shock. Unfortunately, no high-volume manufacturer of regulators is in a position to test for those low noise levels, or to guarantee that you'll never see a noisy part. Please don't expect the manufacturer to admit the parts are bad or unreliable or worthy of being replaced. If you *do* depend on super-quiet ICs, or ICs with other specially selected characteristics, it's wise to keep a spare stock of selected and tested parts in a safe. Then, you can use them when some of the ones you just bought *happen to be* a little too noisy.

What Is Worst Case?

Once I designed a circuit to drive a 200 Ω load (a rather light load) at the far end of a 2000-ft RG174U cable. The specifications called for me to test the circuit by driving the near end of the coaxial cable with a low-impedance square wave. I called the engineer who wrote the spec and recommended that we perform the test with about a 39 Ω source impedance to avoid bad ringing and reflections along the unterminated cable. He told me that this impedance wasn't necessary; he had already checked out the worst-case conditions, with no cable and with 2000 ft of cable. I asked him if he had checked it with 250 ft of cable. "Why, no," he said. So I suggested he try that.

Shortly thereafter, he called me back and agreed that the reflections with 250 ft of cable were intolerable without at least some nominal value of resistance at the source. He had incorrectly assumed that the worst case occurred with the longest cable. It's true that the attenuation was worst with the long length of lossy RG174U cable. But it was this attenuation that caused the ringing and reflections to appear damped out. With the shorter 250-ft cable, a worst-case condition existed at a place he hadn't expected to find it.

So, be cautious about where you look for worst-case conditions. An op amp may exhibit its worst performance at an output voltage other than its maximum negative or positive swing—or even other than zero volts or zero output current. A regulator's worst-case operating conditions may not be at its full-rated load current. When a regulator's power source is resistive, the power dissipation may be higher at three-quarters of its rated current than at full current.

Once I worked on a regulator that ran okay at –55 °C, at room temperature, and at +125 °C, but not at some intermediate temperatures. That was a nasty one. Because some engineers had tested the regulators at hot and cold temperatures and saw no trouble at these extremes, I had to work very hard to convince them not to ship these parts. I had to take them by the hand and show them where the trouble was. It's like a dumb cartoon I once saw showing three men walking out of a movie—an old man, a young man, and a middle-aged man. The posters said the movie was "fun for young and old." And sure enough, the old man and the young man were smiling, but the middle-aged man was frowning. Even a dumb cartoon can be instructive if it reminds you that bad news is not only where you first expect it. It may be lurking in other places, too.

This story reminds me of a boss who asked me if my new regulator design was really short-circuit proof. I told him, yes, I had tested it with short bursts and long pulses and everything in between for days and weeks. With a wry smile, he went over to a tool cabinet and removed a really big, heavy file.[1] He applied this file with rough, uneven scraping motions to ground and to the output of my regulator. He got showers of sparks out of the regulator, but he couldn't kill it. What a "bastard" of a test! Then he explained to me that the random, repetitive action of a file sets up patterns of current loads and thermal stresses that can kill a regulator if its short-circuit protection is marginal. There are many, many tricks you can use to show that a design really can survive every worst-case condition. Every industry has its own tests, and most of them have nothing to do with computers. . . .

Switch-Mode Regulators—A Whole New Ball-Game

These simple tips aren't meant to overshadow the truly difficult areas of regulator design. You might wonder if it's possible for a smart, experienced engineer to design a switch-mode regulator that works well after only minor redesigns and goes into production without a major yield loss. My answer is: *Just barely.* The weasel word here is "smart." If the engineer forgets some little detail and doesn't have a contingency plan to test for it, screen it, or repair the regulators that don't work, then maybe he or she isn't very "smart." Those of us who don't design switchers all the time would have a very poor batting average at getting a design to work right off the drawing board—even if we're really good at designing other circuits. After all, a switcher is a complicated system composed of power transistors, transformers, inductors, one or more control ICs, and lots of other passive components. And, the circuit's layout is critical: The layout must guard small signals against electrostatic interference and cross-talk, and, even more importantly, must control and reject the electromagnetic strays. I mean, for a switcher to be efficient, the volts per microsecond and amperes per microsecond get really *large*, so it doesn't take many picofarads or nanohenries to couple a big noise into the rest of the circuit. The paths for high currents are important, and the paths for cooling air are even more critical.

So when someone asks me how to design a switcher, I ask, "How many do you plan to build?" If the answer is just a few dozen and the design is a full-featured high-power job, I encourage the engineer to buy an existing design. But if large numbers are involved, an engineer usually has the time to do the design right and spread that effort over a few thousand circuits.

An alternative to designing your own switcher is using one of the new "Simple Switchers"—the simple-but-complete switching-regulator ICs. Some of these

1. It was a bastard file.

chips—LM2575, LM2576, LM2577, LM2578—are about as foolproof—for a switcher—as you can get. The data sheets of these parts explain that. You may need a couple resistors, a few capacitors, an inductor, and a fast rectifier, and then it's done. You'll have a cookbook circuit that *really does* work. And if you want to get the component selection information from a program on a floppy disk, I am told that works quite well, too, and is considered pretty "user-friendly." If you only need to supply a few hundred milliamps to your circuit, you may not even need a power transistor or a heat sink. Even in the last couple years there have been advances in designs that really do work, as opposed to "paper designs" that have no chance of surviving a short on the output or of working under worst-case conditions. Although a few of these useless paper designs still pop up from time to time, thank heaven most of them have been driven out.

One of the stories that keeps rattling around the industry is about a group of engineers who decide to band together and start a new computer company. The smartest one is assigned to do the main processor board. Another smart engineer does all the interfaces. And the smart but green "kid engineer" is assigned to do the switch-mode power supply because, of course, that's the easiest part to do. (Anybody who has worked on a big switcher will probably speak up right away: The switcher is *not* as easy as it seems.) In the end, the power-supply design takes a lot longer than everybody expects.

One day, the young engineer opens up the compartment where the balky power supply resides, and it blows up in his face. After his co-workers take the poor fellow to the hospital, they ask around and find a consulting engineer who makes a living out of fixing exactly this kind of switcher problem. The switcher design was slightly off-course and needed the hand of an expert before it would work correctly. So remember, designing switchers is no simple task. Don't hesitate to call in an expert. Note, if this story were not substantially true, the consulting engineer would have starved to death, long ago. I rest my case.

Regulators Suit Different Power Levels

There are several different configurations of switch-mode regulators. At low power levels, capacitively coupled switcher designs are simple, but don't provide much choice of V_{out}: $1.9 \times V_{in}$, $-0.9 \times V_{in}$, and $0.45 \times V_{in}$ are almost the only choices. Flyback regulators are the simplest and cheapest magnetically coupled regulators. However, at power levels above 100 W, their disadvantages become objectionable, and forward or push-pull schemes are more appropriate. At the highest power levels, bridge-type designs are best. If you try to use a configuration at an inappropriate power level, you may have to struggle to get it working. Likewise, the use of current-mode regulation may help you get faster loop response, but the concept is difficult to understand, let alone execute.

Current limiting is always a problem with switchers. The choice of a sense resistor is not easy because the resistor must have low inductance. As with most aspects of switch-mode regulation, to achieve good reliability and to avoid trouble, you have to spend the time to design and test the current-limiting circuit carefully. Some newer switch-mode controller ICs have been engineered to make it reasonably easy. Older ones like the LM3524 haven't been, usually.

Similarly, a soft-start circuit is important for a large switcher, especially when the switcher strains to put out a lot of current to quickly charge up the output filter capacitors, and especially for a boost configuration, where the inductor's current might saturate and refuse to pull the output high enough. For a large supply, this current

parts—units with the start-up circuit broken or disconnected—AND make sure they fail the test. Then leave the test in the flow. Don't drop the test just because nobody has ever seen a part fail. Dropping that test would be courting disaster. Here at National Semiconductor, we've appointed a Czar of Start-up Circuits. He is the repository of all knowledge about circuits that do (and don't) start properly. Since this shy fellow (I shan't give his name) began to reign, the goof-up rate has been cut by many decibels.

References

1. Dietz, Erroll H., "Reduce Noise in Voltage Regulators," *Electronic Design*, Dec. 14, 1989. p. 92.

12. Roundup of "Floobydust"

Loose Ends That Don't Fit Elsewhere

In any serious troubleshooting situation, it is usually wise to plan what tests are most likely to give you the answer quickly, rather than just charging off in random directions. Intermittents are the toughest, most frustrating kind of troubleshooting problem. And bench instruments augment an engineer's senses and open the window of perception to the circuits he or she is troubleshooting. "Floobydust" is an old expression around our lab that means potpourri, catch-all, or miscellaneous. In this chapter, I'll throw into the "floobydust" category a collection of philosophical items, such as advice about planning your troubleshooting, and practical hints about computers and instruments.

Troubleshooting Intermittent Problems

The car that refuses to malfunction when you take it to the shop, the circuit that refuses to fail when you're looking at it—does it really fail only at 2 am?—these are the problems that often require the most extreme efforts to solve.

The following techniques apply to intermittent problems:

1. Look for correlation of the problem with something. Does it correlate with the time of day? The line voltage? The phase of the moon? (Don't laugh.)

2. Get extra observers to help see what else may correlate with the problem. This extra help includes both more people to help you observe and more equipment to monitor more channels of information.

3. Try to make something happen. Applying heat or cold may give you a clue. Adding some vibration or mechanical shock could cause a marginal connection to open permanently, thus leading to identification of the problem and its solution. (Refer to the notes on *The Soul of a New Machine* in Chapter 5).

4. Set up a storage scope or a similar data-acquisition system to trap and save the situation at the instant of the failure. Depending on the nature of the instrument, you may be able to store the data *before* the event's trigger or *after*, or both. This may be especially useful in self-destructive cases.

5. Get one or more buddies to help you analyze the situation. Friends can help propose a failure mode, a scenario, or a new test that may give a clue.

6. As the problem may be extremely difficult, use extreme measures to spot it. Beg or borrow special equipment. Make duplicates of the circuit or equipment that is failing in hopes of finding more examples of the failure. In some cases, you are justified in slightly abusing the equipment in hopes of turning the intermittent problem into an all-the-time problem, which is often easier to solve.

Sugar and Spice and Nothing Nice?

In case you haven't guessed, I'm not a big fan of digital computers and simulation. When a computer tries to simulate an analog circuit, sometimes it does a good job; but when it doesn't, things get very sticky. Part of the problem is that some people put excessive confidence and belief in anything a computer says. Fortunately, my bosses are very skeptical people, and they agree that we must be cautious when a computer makes outrageous promises. Still, we all agree that computers promise some real advantages, if only we can overcome their adversities and problems.

In many cases, if you have trouble with the simulation of an analog circuit or system, you troubleshoot the simulation just as you would the circuit itself. You get voltage maps at various "times" and "temperatures," you insert various stimuli, watch to see what's happening, and modify or tweak the circuit just like a "real" circuit. But, just like the Mario brothers, you can encounter problems in Computerland:

1. You might actually have a bad circuit.

2. You might forget to ask the computer the right question.

3. You might have mistyped a value or instruction or something.

   The easiest mistake of this sort is to try to add a 3.3 M resistor into your circuit. SPICE thinks you mean 3.3 milliohms, not megohms. This problem has hooked almost everybody I know. I solved it by using "3300 K" (3300 kΩ in SPICE), or I may just type out "3300111".

4. You might have a bad "model" for a transistor or device. I've seen a typographical error in the program listing of a transistor's model tie a project in knots for months.

5. You might have neglected to include strays such as substrate capacitance, PC-board capacitance, or—something that most people forget—lead inductance.

6. You might get a failure to converge or an excessive run time. Or the computer might balk because the program is taking too many iterations.

   Sometimes problems happen that only a computer expert can address. But when you ask the computer guru for advice, you might get no advice or—what's worse— bad advice. After all, many computer wizards know nothing whatsoever about linear circuits. If the wizard tells me, "Hey, don't worry about that," or, "Just change the voltage resolution from 0.1 mV to 10 mV," then I must explain to the wizard that, although that advice might make some computers happy, it gives me results that are completely useless. Talking to computer wizards is sometimes difficult.

   Even if you do everything right, the computer can lie to you. Then you have to make a test to prove that you can get the right answer and the computer can't.

   For example, one time we had a circuit with 60 transistors, and a diff-amp appeared to be oscillating even though it was clearly switched OFF. The computer experts told us that we had obviously made a mistake. So we disconnected and re- moved 58 of the transistors—there was nothing left but 2 transistors, and one of them was biased off by a full volt. And its collector current was "oscillating" at 100 kHz, between plus *and* minus 10 μA, even though nothing in the circuit was moving. When we confronted the computer experts, they belatedly admitted there was an "internal timing error," which they proceeded to fix. But it certainly took us a diffi- cult week to get them to admit that.

Photo Courtesy Al Neves.

Figure 12.1. I hurled this computer to its doom from atop National's 3-story parking garage. As the dust settled, I knew *that* computer would never lie to me again!

My boss tells me that I should not be so negative, that computers are a big part of our future. When he says that, I am tempted to go out and buy stock in companies that make Excedrin™ and antacids, because that also will be a big part of our future . . .

I have given a couple lectures at major conferences, with comments about SPICE and some of its problems. (Ref. 1 and 2). After the lecture, engineers from other companies have come up and told me, "Yeah, we have those kinds of problems, too. . . ." (For additional comments on SPICE, refer to Appendix G.)

One guy gave me a tip: "Don't put a 50-Ω resistor in your circuit—put in 50.1 Ω, and it may converge better." In other cases, we discovered that a parallel resistor-capacitor combination that was connected only to ground was helping to give us convergence; when we "* commented out" the R and C, we could not get convergence any more. Other people comment that if you change the name of a resistor, or its number, or its position on the list of components, convergence may be improved—or ruined. So this convergence bird is a very fragile and flighty thing.

My boss reminds me that some versions of SPICE are better than others at converging, and I shouldn't just be a complainer. But I am just reminding you that all kinds of computer simulation get criticized, and sometimes the criticism is valid—the complainer is not just imagining things (Ref. 3).

So if the computer persists in lying to you, just tell your boss that the computer has proved itself incompetent. Junk that digital piece of disaster!!

What I really think you ought to do, instead of using digital simulations, is to make an analog-computer model—you'll have a lot less trouble. Be sure to scale all the transistors' capacitances at 100× or 1000× their normal values, so the time scale is scaled down by 100×, which makes the strays negligible. That's what I do. I have seen it work when SPICE cannot be beaten into cooperation. You might call it an "analog computer," because that's exactly what it is. I will listen to alternative points of view but, be forewarned, with frosty skepticism.

Lies, Damned Lies, and Statistics

One thing that doesn't help me a darned bit is "statistics," at least in the sense that mathematicians use them. I find most statistical analyses worse than useless. What I *do* like to use is charts and graphs. The data I took of the diodes' V_F versus I_F back in Chapter 6 were a little suspicious when I wrote down the numbers, but after I plotted the data, I *knew* there was something wrong. Then I just went back and took more data until I understood what the error was—AC current noise crashing into my experiment, causing rectification. If data arise from a well-behaved phenomenon and conform to a nice Gaussian distribution, then I don't care if people use their statistical analyses—it may not do a *lot* of harm. (Personally I think it does harm, because

Photo Courtesy Fran Hoffart.

Figure 12.2. SPICE printouts are almost always good for *something*.

when you use the computer and rely on it like a crutch, you get used to believing it, and trusting it without thinking. . . .) However, when the data get screwy, classical statistical analysis is worse than useless.

For example, one time a test engineer came to me with a big formal report. Of course, it didn't help things any that it arrived at 1:05 for a Production Release Meeting that was supposed to start at 1:00. But this was not just any hand-scrawled report. It was handsome, neat, and computerized; it looked professional and compelling. The test engineer quoted many statistical items to show that his test system and statistical software were great, even if the ICs weren't. Finally he turned to the last page and explained that, according to the statistics, the ICs' outputs were completely incompetent, way out of spec, and thus the part could not be released. In fact, he observed, the median level of the output was 9 V, which was pretty absurd for the logical output of an LM1525-type switching regulator, which could only go to the LOW level of 0.2 V or the HIGH level of 18.4 V. How could the outputs have a median level of 9 V?? How do you get an R-S flip-flop to hang up at an output level halfway between the rails? Unlikely. . . . Then he pointed out some other statistics—the 3 σ values of the output were +30 V and –8 V. Now, that is pretty bizarre for a circuit that has only a +20-V supply and ground (and it is not running as a switching regulator—it's just sitting there at DC). The meeting broke up before I could find the facts and protest, so that product was not released on schedule.

It turned out, of course, that the tester was running falsely, so while the outputs were all supposed to be SET to +18.4 V, it was actually in a random state, so that half the time the outputs were at 18.4 V and half the time at 0.2 V. If you feed this data into a statistical program, it might indeed tell you that some of the outputs might be at 9 V, assuming that the data came from a Gaussian distribution. But if you *look* at the data and *think*, it is obvious that the data came from a ridiculous situation. Rather than just try to ram the data into a statistical format, the engineer should start checking his tester.

Unfortunately, this engineer had so much confidence in his statistical program that he spent a whole week preparing the Beautiful Report. Did he go report to the design engineer that there were some problems? No. Did he check his data, check the tester? No. He just kept his computer cranking along, because he knew the computer analysis was the most important thing.

We did finally get the tester fixed, and we got the product out a little late, but obviously I was not a fan of that test engineer (nor his statistics) as long as he was at our company. And that is just one of a number of examples I trot out, when anybody tries to use statistics when they are inappropriate.

I *do* like to use scatter plots in two dimensions, to help me look for trends, and to look for "sports" that run against the trend. I don't look at a lot of data on good parts or good runs, but I study the *hell* out of bad parts and bad runs. And when I work with other test engineers who have computer programs that facilitate these plots, I support and encourage those guys to use those programs, and to *look at* their data, and to *think about* those data. Anything that facilitates thinking—*that* I support.

Keep It Cool, Fool....

A couple years ago I was approached by an engineer who was trying to use one of our good voltage references that had a typical characteristic of about 20 ppm per 1000 hours long-term stability at +125 °C. He was using it around room temperature, and he was furious because he expected it to drift about 0.1 ppm per 1000 hours at room temp, and it was a lot worse than that. Why was our reference no good, he asked? I pointed out that amplifiers' drifts and references' drifts do *not* keep improving by a factor of 2, every time you cool them off another 11 degrees more. I'm not sure who led him to believe that, but in general, modern electronic components are not greatly improved by cooling or the absence of heating. In fact, those of us who remember the old vacuum-tube days remember that a good scope or voltmeter had an advantage if you kept it running nice and warm all the time, because all the resistors and components stayed dry and never got moist under humid conditions.

I won't say that the electrolytic capacitors might not have liked being a little cooler. But the mindless effort to improve the reliability by keeping components as cool as possible has been overdone. I'm sure you can blame a lot of that foolishness on MIL-HBDK-217 and all its versions. In some businesses you have to conform to -217, no matter how silly it is, but in the industrial and instrument business, we don't really have to follow its every silly quirk and whim. One guy who is arguing strenuously about -217 is Charles Leonard of Boeing, and you may well enjoy his writing (Ref. 4). So if something is drifting a little and you think you can make a big improvement by adding a fan and knocking its temperature down from +75 to +55 °C, I'm cautioning you, you'll probably be disappointed because there's not usually a lot of improvement to be had. It is conceivable that if you have a bad thermal pattern causing lots of gradients and convection, you can cut down that kind of thermal problem, but in general, there's not much to be gained unless parts are getting up near their max rated temperature or above +100 degrees. Even plastic parts can be reliable at +100 degrees. The ones I'm familiar with are.

There's Nothing Like an Analog Meter

Everybody knows that analog meters aren't as accurate as digital meters. Except... you can buy DVMs with a 0.8% accuracy; analog meters better than that exist. Anyway, let's detail some problems with analog meters.

Even if an analog meter is accurately calibrated at full scale, it may be less accurate at smaller signals because of nonlinearity arising from the meter's inherent imperfections in its magnetic "circuits." You can beat that problem by making your own scale to correct for those nonlinearities. Then there's the problem of friction and hysteresis. The better meters have a "taut-band" suspension, which has negligible friction—but most cheap meters don't. Now, as we have all learned, you can neutralize most of the effects of friction by gently rapping on, tapping at, or vibrating the meter. It's a pain in the neck, but when you're desperate, it's good to know.

Even if you don't shake, rattle, or roll your meters, you should be aware that they are position-sensitive and can give a different reading if flat or upright or turned sideways. The worst part about analog meters is that if you drop them, any of these imperfections may greatly increase until the meter is nearly useless or dead. This is "position sensitivity" carried to an extreme. Ideally, you would use digital meters for every purpose. But analog meters have advantages, for example, when you have to look at a trend or watch for a derivative or an amplitude peak—especially in the presence of noise, which may clutter up the readings of a digital voltmeter. So, analog meters will be with us for a long time, especially in view of their need for no extra power supply, their isolation, and their low cost.

But, beware of the impedance of meter movements. They look like a stalled motor—a few hundred millihenries—at high frequencies. However if the needle starts swinging, you'll get an inductive kick of many henries. So, if you put an analog meter in the feedback path of an op amp, you'll need a moderate feedback capacitor across the meter.

Digital Meters—Not So Bad, and Sometimes Better than That

As I mentioned before, digital meters are always more accurate than analog meters...except for when they aren't. Recently, a manufacturer of power supplies decided to "modernize" its bench-type supplies by replacing the old analog meters with digital meters. Unfortunately, these meters came with an accuracy of ±5%. Having a 2-1/2-digit digital panel meter (DPM) with a resolution of 1 part in 200 but an accuracy of 1 part in 20 certainly is silly. Needless to say, I stopped buying power supplies from that manufacturer.

The steadiness and irrefutability of those glowing, unwavering digits is psychologically hard to rebut. I classify the readings of the DVM or DPM with any other computer's output: You have to learn to trust a computer or instrument when it's telling the truth, and to blow the whistle on it when it starts to tell something other than the truth.

For example, most slow DVMs have some kind of dual-slope or integrating conversion, so they're inherently quite linear, perhaps within 1 or 2 least-significant digits. Other DVMs claim to have the advantage of higher conversion speed; this higher speed may be of no use to the bench engineer, but it is usable when the DVM is part of an automated data-acquisition system. These faster instruments usually use a successive-approximation or recirculating-remainder conversion scheme, both of which are not inherently linear but depend on well-trimmed components for linearity. I have seen several DVMs that cost more than $1000 and were prejudiced against certain readings. One didn't like to convert 15 mV; it preferred to indicate 14 or 16.

One time I got a call from an engineer at one of the major instrument companies. He wondered why the Voltage-to-Frequency converter he made with an NSC LM331 was showing him poor linearity—worse than the guaranteed spec of 0.01%. I told him that was strange, because if it was true, it was the first LM331 to have poor linearity from the first couple million we had produced. I advised him to check the capacitors and the op-amp waveform, and to call me back, because if he had a part that didn't meet specs, I wanted to get my hands on it.

The next day he called me back, feeling very sheepish and embarrassed. He admitted he had been using a prototype DVM designed by his company, and because it was a prototype, it was not exactly under calibration control. It was his DVM that had gone out of linearity, *not* the LM331.

Normally I hate to use a DVM's autoranging mode. I have seen at least two (otherwise high-performance) DVMs that could not lock out the autorange feature. The worst aspect of these meters was that I couldn't tell where they would autorange from one range to another, so I couldn't tell where to look for their nonlinearity; yet I *knew* there was some nonlinearity in there somewhere. After an hour of searching, I found a couple of missing codes at some such preposterous place as 10.18577 V. And this on a $4000 DVM that the manufacturer claimed could not possibly have such an error—could not have more than 1 ppm of nonlinearity.

Another fancy DVM had the ability to display its own guaranteed maximum error, saying that its own error could not be more than $\pm 0.0040\%$ when measuring a 1 MΩ resistor. But then it started indicating that one of my better 1.000000 MΩ resistors was really 0.99980 MΩ. How could I prove if it was lying to me? Easy—I used jiu-jitsu—I employed its own force against itself. I got ten resistors each measuring exactly 100.000 kΩ—the fancy machine and all the other DVMs in the lab agreed quite well on these resistors' values. When I put all 10 resistors in series, all the other meters in the lab agreed that they added up to 1.00000 MΩ; the fancy but erroneous machine said 0.99980 MΩ. Back to the manufacturer it went.

So, if you get in an argument with a digital meter, don't think that *you* must necessarily be wrong. You can usually get an opinion from another instrument to help prove where the truth lies. Don't automatically believe that a piece of "data" *must* be correct *just* because it's "digital."

And be sure to hold onto the user's manual that comes with the instrument. It can tell you where the guaranteed error band of the DVM gets relatively bad, such as for very low resistances, for very high resistances, for low AC voltages, and for low or high frequencies....

Most digital voltmeters have a very high input impedance (10,000 MΩ typ) for small signals. However, if you let the DVM autorange, at some level the meter will automatically change to a higher range where the input impedance becomes 10 MΩ. Some DVMs change at ± 2 V or 3 V, others at 10 or 12 or 15 V, and yet others at ± 20 V. As I mentioned in the chapter on equipment, I like to work with the DVMs that stay high-impedance up to at least 15 V. But, the important thing is to know the voltage at which the impedance changes. A friend reminded me that his technician had recently taken a week's worth of data that had to be retaken because he neglected to allow for the change of impedance. I think I'll go around our lab and put labels on each DVM.

Still, DVMs are very powerful and useful instruments, often with excellent accuracy and tremendous linearity and resolution—often as good as 1 ppm. I've counted some of these ultralinear meters as my friends for many years. I really do like machines —such as the HP3455, HP3456, and HP3457—that are inherently, repeatably linear, as some of these DVMs are absolutely first-class.

One picky little detail: Even the best DVM is still subject to the adage, "Heat is the enemy of precision." For example, some DVMs have a few extra microvolts of warm-up drift, but *only* when you stand the box on its end or side. Some of them have a few microvolts of thermal wobble and wander when connected to a zero-volt signal (shorted leads), but only when you use banana plugs or heavy-gauge (16, 18, or 20 gauge) leads—not when you use fine wire (26 or 28 gauge). The fine-wired leads apparently do not draw as much heat from the front-panel binding posts. So, even the best DVM auto-zero circuit cannot correct for drifts outside its domain.

Most engineers know that DVMs add a resistive (10 MΩ) load to your circuit and a capacitive load (50 to 1000 pF) that may cause your circuit to oscillate. But, what's not as well known is that even the better DVMs may pump noise back through their input terminals and spray a little clock noise around your lab. So if you have a sensitive circuit that seems to be picking up a lot of noise from somewhere, turn off your DVM for a few seconds to see if the DVM is the culprit. If that's not it, turn off the function generator or the soldering iron. If it is the DVM's fault, you may want to add RC filters, RLC filters, or active filter/buffers with precision operational amplifiers, to cut down on the noise being injected into your circuit. There is a little RC filter shown in Figure 2.4 of Chapter 2, that is useful for keeping the noises of the DVM from kicking back. Or, you might want to go to an analog meter, which—as we discussed on a previous page—do not have any tendency to oscillate or put out noise. An analog meter with a battery-powered preamplifier will not generate much noise at all, by comparison to a DVM....

Signal Sources

While I'm on the subject of instruments, I really enjoy using a good function generator to put out sines and triangle waves and square waves and pulses. I love my old Wavetek 191. But I certainly don't expect the signals to be absolutely undistorted—all these waveforms will distort a little, especially at high frequencies. So if I want my function generator to give me a clean sine wave, I put its output through an active filter at low frequencies or an LC filter at high frequencies. If I want a clean, crisp square wave, I will put the signal through a clipping amplifier or into a diode-limited attenuator (Figure 12.3). If I want a cleaner triangle than the function generator will give me, I just make a triangle generator from scratch.

But a function generator lets me down when some absent-minded person pushes one button too many and the output stops. (Usually, that absent-minded person is *me*.) It can take me five minutes to find what the problems are. I love all those powerful, versatile functions when I need them, but they drive me nuts when the wrong button gets pushed.

Similarly, a scope's trace can get lost and hide in the corner and sulk for many minutes on end if you don't realize that somebody (maybe your very own errant fat finger) pushed a treacherous button. When the digital scopes with their multiple layers of menus and submenus start playing that game, I find I need a buddy system—somebody to come and bail me out when I get hopelessly stuck. What menu is that dratted beamfinder on, anyway?

But, scopes work awfully well these days. Just don't expect precision results after you drive the trace many centimeters off scale by turning up the gain to look at the bottom of a tall square wave. Most scopes aren't obligated to do that very well. Similarly, be sure to keep the trimmers on your 10× probes well adjusted, and run a short ground path to your probes when you want to look at fast signals, as discussed in Chapter 2.

Troubleshoot As You Go

Some people like to build up a big system and turn on the power; and, Voila, it doesn't work. Then they have to figure out what kind of things are wrong in the

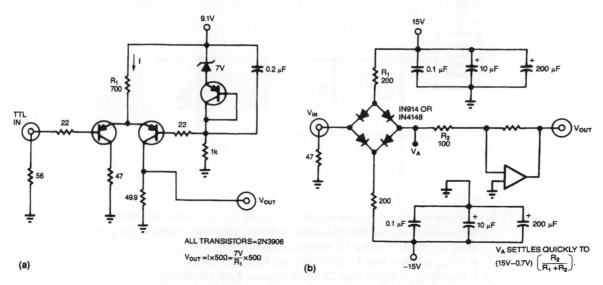

Figure 12.3. Either a clipping amplifier (a) or diode-limited bridge (b) will give you a clean, crisp square wave.

whole megillah. I prefer to build up modular chunks, and to test each section as I get it built. Then if it works, that provides a pleasant positive kick, several times along the length of a project. *But* if it doesn't work, it gives me a chance to get it on track before I go charging ahead and get the whole thing finished. Sometimes it's just a missing capacitor. Other times, I've got the whole concept wrong, and the sooner I find it out, the better. So if you see one of my systems made up of 14 little 7-inch square sections, all lashed together on a master framework, don't be surprised. I mean, if *you* can make a big system work the first time, more power to you. I often remind my technician, "This may not work the first time, but it will be *really close.* You may have to tweak an R here or a C there, but it won't be disastrously bad."

Similarly, when I have a circuit that does not work right—do I just want to get it working right? Rather not. What I want is to learn what was wrong, and learn what happens when I try changes. So I don't give my technician a long list of changes to make, all at once. I tell him, make this change first and see if the gain gets better. If that doesn't work, make *that* change and then *that* one, and keep an eye on the gain and the phase. Then try *this* tweak on the output stage, and trim it for lower distortion at 10 kHz.... If he made all the changes at once, the performance might improve, but if we weren't sure which changes made the improvement, we wouldn't be learning much, would we?

Systems and Circuits

When a system is designed, it is usually partitioned into subsections that are assigned to different people or groups to engineer. Two very important ingredients in such a system are Planning and Communication. If the partitioning was done unfairly, then some parts of the system might be excessively easy to design, and other parts substantially impossible. We've all seen that happen, so we must be careful to prevent it from happening to *our* systems. For if all the subsystems work except for *one,* the whole project will probably fail.

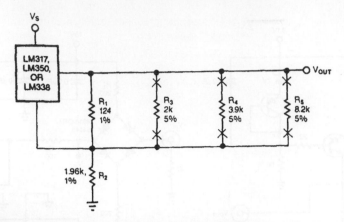

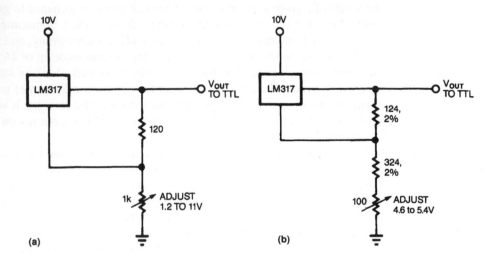

Figure 12.4. If you're worried that some foolish person will ruin a circuit by misadjusting a trimming potentiometer, you can foil the bungler with this "snip-trim" network. The procedure for trimming V_{out} to 22 V within 1% tolerance is as follows:
- If V_{out} is higher than 23.080 V, snip out R_3 (if not, don't);
- then if V_{out} is higher than 22.470 V, snip out R_4 (if not, don't);
- then if V_{out} is higher than 22.160 V, snip out R_5 (if not, don't).

Obviously, you can adapt this scheme to almost any output voltage. Choosing the breakpoints and resistor values is only a little bit tricky.

Figure 12.5. One of my pet peeves: an excessive trimming-potentiometer adjustment range (a). The circuit in (b) suits TTL much better.

The need for good communications is critical—good oral *and* written communications, to prevent confusion or false assumptions. After all, it's not realistic to expect the system design and every one of its definitions to start out perfect from the first day. The chief troubleshooter for the system should be the Program Manager, or whatever he is called. He'll have his computers, Pert charts, GANTT charts, and so forth, but, most valuable, he has his people who must be alert for the signs of trouble.

These people have to be able to communicate the early signs of trouble, so the leader can get things fixed. (Well, OK, he *or* she....)

How to Trim without Trimming Potentiometers

Speaking of keeping circuits well trimmed, some people like to use trimming potentiometers to get a circuit trimmed "just right." Other people hate to, because the potentiometers are expensive or unreliable or drifty. Worst of all, if a circuit can be trimmed, it can also be mis-trimmed; some person may absentmindedly or misguidedly turn the potentiometer to one end of its range or to the wrong setting. How long will it take before that error is corrected?

For just this reason, some people prefer fixed-voltage regulators because they always have a valid output ($\pm 5\%$) and can never get goofed up by a trimming potentiometer. Other people need a tighter tolerance yet are nervous about the trimming potentiometer. You will find the solution in the snip-trim network in Figure 12.4. (Ref. 5). This scheme will let you trim a regulator well within 1% without trimming potentiometers. Note that you could also use this technique to set the gain of integrators and the offset of amplifiers. It's not always easy to engineer the correct values for these trims, but it is possible. And, nobody's going to go back and tweak the potentiometer and cause trouble if there's no potentiometer there to tweak.

A pet gripe of mine concerns engineers who design a circuit with an adjust range that's so wide that damage can occur. For example, Figure 12.5a is a bad idea for a regulator for a 5-V logic supply because the TTL parts would be damaged if someone tweaks the pot to one end of its range. Figure 12.5b is better.

What about Solderless Breadboards

Here's a chunk of late-breaking floobydust—the topic is those solderless breadboards, which consist of a number of metal strips and solderless connectors hidden underneath a plastic panel with lots of holes in it. Schools often use them to introduce students to the joys of breadboarding because you can easily connect things by just stuffing wires and components into the holes. The problems begin with capacitance. The breadboards usually have 2, 3, or 5 pF between adjacent strips. On a good day, only a wise engineer could plan a layout that all the capacitors, sprinkled throughout the circuit, wouldn't ruin.

The next serious problem with solderless breadboards is the long leads, which make adding effective power-supply bypass capacitors close to a chip difficult.

Next, I suspect that some of these panels, although they are not inexpensive, use cheap plastics such as nylon. On a warm, humid day, cheap plastics do not offer high insulation resistance. Nobody wants to talk about what kind of plastic the breadboards are made of.

Finally, Mr. Scott Bowman of Dublin, CA, points out that after you insert enough wires into any given hole, the solderless connector will scrape sufficient solder off the wire so that the scraps of solder will pile up and start to intermittently short out to an adjacent strip. Further, the adhesive that holds on the back panel tends to hold the solder scraps in place, so you can't clean the scraps out with a solvent or a blast of air.

I didn't even think about these solderless breadboards when I wrote my series because I see them so rarely at work. They just have too many disadvantages to be good for any serious work. So, if you insist on using these slabs of trouble, you can't say I didn't warn you.

References

1. Robert A. Pease, *Practical Considerations for the Design of High-Volume Linear ICs*, IEEE International Symposium for Circuits and Systems, April 1990.

2. Robert A. Pease, "Band-Gap Reference Circuits: Trials and Tribulations," IEEE Bipolar Circuits and Technology Meeting, September 1990.

3. Robert A. Pease, "What's all this SPICEy Stuff Anyhow?" *Electronic Design*, December 1990.

4. Leonard, Charles, "Is reliability prediction methodology for the birds?," *Power Conversion and Intelligent Motion*, November 1988, p. 4.

5. Robert A. Pease, "A New Production Technique for Trimming Voltage Regulators," *Electronics*, May 10, 1979, p. 141. (Also available as Linear Brief LB-46 in the *Linear Applications Databook*, National Semiconductor Corp., 1980–1990)

13. Letters to Bob

My series in *EDN* on troubleshooting generated myriad letters from readers. Because so many of the letters contained worthwhile troubleshooting tips and amusing personal anecdotes, we decided to collect some of the best letters into a chapter, along with my replies and interjections. The tips just keep on coming. . .

Dear Bob:

Here are some tips and gotchas:

1. A significant source of noise in my lab is the ever-present video-display terminal. It couples especially well to audio-frequency transformers.

2. I head off a lot of trouble by providing RF bypass on audio and DC circuits. Their audio and low-frequency-only inputs can pick up AM radio. Having music come out of a speaker that is supposed to be a monitor on a telephone circuit is very bad form.

3. My computer brethren frequently fail to consider what happens during reset. I saw a thermal printer catch on fire once when its internal μP was reset. The reset 3-stated the printhead's drivers, which allowed all of them to turn on continuously. Later a software bug turned them on continuously again. I finally made the printhead computer-proof by capacitively coupling the drivers so that the μP had to produce a continuous sequence of pulses to keep the heads turned on.

4. My inexperienced brethren frequently forget to calculate total power supply requirements.

5. Vishay (Malvern, PA) produces some very accurate, very stable (0.6 ppm/°C) resistors, which I keep around to check ohmmeters.

6. Some companies think they are helping designers by taking instruments to the calibration lab without letting the designers know. They do not understand that small day-to-day drifts are less annoying than an unexpected step change produced by recalibration.

7. My computer brethren frequently lose scope-probe ground clips—the clips get in their way and sometimes short a power supply. I gave up and bought a pile of the clips that I keep in my bench.

8. Sad but true, sometimes adding a scope probe to a malfunctioning circuit makes the circuit work. The probe adds enough capacitance to kill a glitch or stop a race. On floating CMOS, the DC impedance of the scope can be low enough to pull the signal down to a valid level.

9. At one place I worked, I was called to the factory to make my "no-good" circuit work. The complaint was that the DC offset of an op amp was drifting. When I got there, I found the technician had a good DVM connected to the op amp through a piece of coaxial cable to keep out noise. Of course, the cable's capacitance was making the op amp oscillate. You can't measure DC parameters when the op amp is oscillating. Sometimes I find a scope connected this way because the tech wanted more gain or could not find a 10× probe.

10. Probes work into a certain, specified scope-input capacitance. You can't always take a probe that came with one scope and use it at high frequencies on another scope.

11. A simple test technique is waving your hand over a circuit to feel for the hot spot. If something has gone into a latch-up but is not smoking, you can frequently find it this way.

12. Edmund Scientific (Barrington, NJ) sells thermally sensitive liquid-crystal sheets, which you can lay over a circuit to find moderate hot spots. This material works well when you have a known-good PC board to compare with the circuit under test.

13. Drafting departments sometimes erroneously think that they own the schematic and that its only purpose is to serve as a wiring diagram for the PC-board layout. Long after PC-board layout, the production-test, sustaining-engineering, and service departments will still need the schematic. Drafting tends to lose notes that I place on the schematic, such as filter poles and zeros, temperature coefficients, normal AC and DC voltages, waveforms, and thermal information. I save myself a lot of calls by putting this information in front of the techs from the beginning.

14. You can make an extremely low-distortion (and slightly microphonic) sine-wave oscillator from a light bulb and an op amp. I got the circuit from Linear Technology's (Milpitas, CA) AN 5 application note. I built a 3-frequency (400-, 1000-, and 2800-Hz) oscillator in a small metal Bud box. It had a THD lower than –80 dB.

15. If a circuit's DC values change when you breathe on it, you may have dirty circuit boards.

16. When testing high-gain, low-signal-level circuits, repeat the measurements with the lights off. You may be surprised to learn that many components are photosensitive and have infrared transparent bodies. One of my colleagues had a photosensitive metal-can op amp that leaked light in around the leads.

17. Protection diodes can rectify high-frequency noise and oscillations.

18. Micro Technical Industries (Laguna Hills, CA) makes a handy thermal probe with which you can individually heat components. The probe has tips to fit various components, such as small and large resistors, metal-can op amps, and DIPs of various sizes.

19. Some sample-and-hold circuits are sensitive to slew rates on the digital inputs.

20. Even Schmitt triggers can exhibit metastability.

21. As paraphrased from an Analog Devices (Norwood, MA) application note, "You may be able to trust your mother, but you should never trust your ground."

22. Wrapped-wire circuits work pretty well if you can distribute power and ground properly. I use large-diameter bus wire in a rectangular grid for high-frequency logic if I don't have a wrapped-wire board with internal power and ground distribution.

23. Sometimes powering your test circuit with batteries breaks ground loops and eliminates power-line noise.

24. A handy thing to have is a 60-Hz, passive, twin-T notch filter in a small Pomona box with dual hanana plugs for input and output.

25. Another handy thing to have is a 20-dB high-impedance amplifier in another Pomona box. The circuit in Figure 13.1 works at audio frequencies.

26. The CMRR of an op amp is not a constant function of the common-mode voltage. This inconsistency often dominates nonlinearity in noninverting circuits.

27. Getting some engineers to hold design reviews is hard. By law, our drafting department will not start a PC-board design until the designer hosts a review. The drafting
supervisor can easily enforce this policy because he does not report to the first-level
engineering managers. The moral indignation of the designer's peers informs the
quality of the review.

28. Those partial but detailed schematics National Semiconductor sometimes places on
data sheets provide valuable insight into how a part may be acting in unusual circumstances. Please encourage National to continue the practice.

Roy McCammon
3M/Dynatel
Austin, TX

Hello to Mr. McCammon:

Well, I thank you for your comments, many of which are excellent. I will comment
on some of them individually, but, as a collection, they are the best new ideas that
anybody has given me.

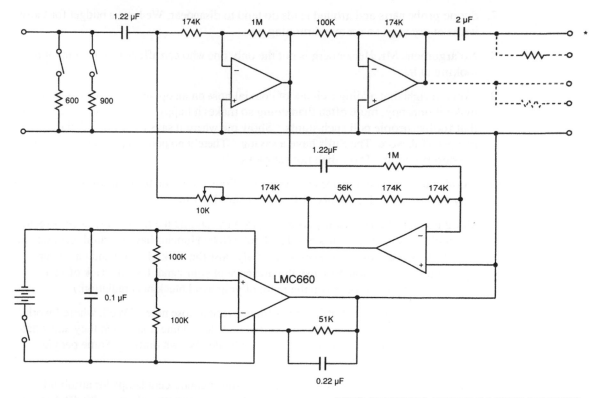

* RAP SUGGESTS ADDING DASHED CIRCUITS

Figure 13.1. This 20-db high-impedance amplifier works at audio frequencies and is a handy
troubleshooting aid. I suggest adding the dashed circuits as useful options.

1. Noise from a video terminal? I have not seen that problem, but such noise may be a serious problem in some cases. I don't have a *digital* computer near my workbench, but other people may. As I mentioned in Chapter 2, if you bring an AM radio near a computer or keyboard, the radio will detect various amounts of graunchy RF noises.

2. I, too, recommend RF bypassing. But you point out that bypassing should be *against* ambient RF. I rarely think of bypassing in those terms, but you are right.

3. Yeah, I believe that. Some people are capable of building circuits that are the exact opposite of "fail-safe."

4. Yeah, I believe that too. But data sheets for digital as well as linear ICs do not help because they only indicate quiescent power drain, and give no clue about what will change when the output is swinging slow or fast. Even TTL draws more current when running fast.

5. I'm sure that most Vishay resistors are quite stable; I myself keep a group of old wire-wounds for that purpose. Some of them are so old, they are older than the oldest Vishays. Consequently, I have confidence that they have good long-term stability, which I find more important than temperature coefficient.

6. I agree that spiriting away equipment for calibration is a serious issue. In our group, we used to have equipment disappear for calibration just when we needed the equipment desperately. We finally resolved the problem by requiring that engineers put their equipment on a calibration shelf when it was due to be calibrated. If they don't put it there, the calibration guys won't steal it.

7. Scope probe parts and ground leads do tend to disappear. We have a budget for those items every year, so we avoid running out.

8. No argument. Mr. Heisenberg is not the only one who can affect a measurement by looking at it.

9. You are right that adding a chunk of coaxial hose on an op-amp's output can sure make it unhappy, more often than doing so makes it happy. You are right to complain that foolish people pull such stunts. "Show me where it says I can't do it," they protest. (Ed. Note: The Sufis have a saying, "There's no point in putting up signs in the desert saying, 'Thou shalt not eat rocks.'")

10. Good statement about probe compatibility. Sometimes you turn the adjustment all the way to the end, and the probe still won't neutralize out.

11. Good point. I often check my soldering iron by running the tip past my nose—about one inch away—to see if it's good and hot. (Note: Humans have infrared receptors in their lips. If you close your eyes and slowly raise the back of your hand past your lips, you should be able to sense the presence of your hand. Use the back of your hand because the calls on the palm of your hand block heat radiation.)

13. You point out that ". . . drafting departments tend to lose notes." Well, where I work, drafting departments do a task, and if that includes adding notes, then they add whatever the engineer requires. Actually, I usually do my own drafting. Some people gripe, but the information is all there.

14. I don't usually get too enthusiastic about using incandescent lamps for amplitude control. It's true that most oscillators don't have distortion as low as −80 dB, but you can put the output of a mediocre oscillator through a filter and get distortion lower than −80 dB.

16. Yes, I believe that. We have found that our lowest input current amplifiers such as

LMC662 have the lowest leakages in plastic DIPs, which are made by an automated process and are untouched by human hands. The TO-99 and CERDIP packages are not nearly as good as the plastic minidips which do 10^{15} Ω repeatably.

17. "Protection diodes can rectify high-frequency noise." I have never seen this one. Boy, Roy, you must live in a *nasty* neighborhood for ambient RF noises. You can probably run a transistor radio without any batteries.

18. I forgot to even mention thermal probes. We use such probes more for characterizing than for troubleshooting. Often, a soldering iron does the troubleshooting job faster but more crudely.

19. The LF198 data sheet mentions that you should not let the Sample input move too slowly. Are there other S/H circuits that are touchy? Ones that do not mention this fact in their data sheets?

22. I rarely work with wrapped-wire stuff, but I bet a lot of people get fooled by bad daisy-chaining of power-supply runs and lousy power-supply bypassing—whether for linear or digital ICs.

23. I rarely find batteries necessary, but, in extreme cases, they are useful.

24. I rarely find notching out 60- or 120-Hz interference necessary. I usually subtract the 60-Hz noises visually from a scope trace.

25. Yes, portable preamps are often useful.

26. Just as I was saying back in Chapter 8—it's silly to assume that the CM error is linear.

27. Design reviews are a good idea. But even if the circuit design is perfect, I find the lay-out to be pretty critical. So a beer check by all your buddies is awfully important, too.

RAP

Dear Bob:

As a practicing technician for many years, I want to comment on one or two things I read in your series and perhaps pass on an experience or two.

On page 130 of the August 17, 1989 article (new Chapter 7) you mention the possibly harmful side effects of drawing base current out of a transistor. Most (if not all) of the switched-mode power supplies I have come across appear to do just that—to switch the transistor off more quickly by removing carriers from the base.

This technique appears to work well in practice. I have used the technique in many of the inverters I have designed, and (on the face of it, at least) there doesn't appear to be any component deterioration over time. I usually use some form of reverse-voltage limiting to ensure that the base-emitter junction doesn't undergo zener break-down.

Referring to your comments on p. 132 of the same article, I have to disagree on the advisability of plugging MOS ICs into sockets with the power applied. I consider it inadvisable because power may easily be sourced to the chip via its input and output pins in this situation. Some manufacturers forbid this by implication in the "Electrical Characteristics" section of their data sheets, and I have witnessed device destruction being caused by this practice. I am also unable to agree with your comments on not

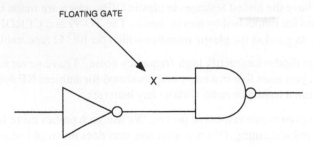

Figure 13.2. Letting TTL inputs float HIGH is a risky business. They won't do any harm on your bread-board—only when you go into production. It's wise to tie them up towards $+V_S$ with 1 kΩ or so.

wearing ground straps when handling MOS ICs. My experience, and that of others, has been that MTBFs plummet if you don't handle MOS ICs with kid gloves. The problem is that the device rarely fails on the spot, but having possibly been over-stressed, will do so at a time and place of its own choosing.

Turning to the article in the September 28, 1989, edition of EDN (Chapter 4), I feel bound to present a caution regarding tantalum capacitors. First, they are even less tolerant of reverse polarity than electrolytic capacitors are. Reverse polarity can arise when, say, coupling op amps with tantalum capacitors. Second, I have replaced more tantalum capacitors than I can remember because they short-circuited for no good reason.

The worst place for a bad tantalum capacitor is on a computer motherboard, which is just where I found the last one I replaced. The job started out as a short that was shutting down the power supply. Ascertaining which rail the problem was on was easy enough. I decided that because of the number of components on the bus, I would have to try something less radical than temporarily isolating a section of the rail. I tackled the problem by feeding a 10-kHz sine wave at about 1 V rms through a 1 kΩ resistor and monitoring the traces with a current probe. I found the offending compo-nent in less than a minute. I may have been lucky, but, on the other hand, my test signal wouldn't cause the ICs to draw any significant current and was of a low-enough frequency not to cause enormous current in the good decoupling capacitors.

In the same article, you said that floating TTL inputs is reasonably OK. In some situations, particularly noisy ones, I would have to disagree. I have seen nasty prob-lems where this practice causes random glitches. The circuit in Figure 13.2 is an absolute nightmare. Yet this practice is common on IBM PC clones from Asian man-ufacturers in, of all things, the glue logic of 80386 motherboards, where high-speed clocks are the norm.

Malcolm Watts
Wellington Polytechnic
Wellington, New Zealand

Dear Mr. Watts:

Thank you for your comments. You question the practice of pulling current out of a transistor's base circuit. If you were to actually pull current out of the base and

Figure 13.3. This scheme will make your tantalum capacitors immune to damage from reversal.

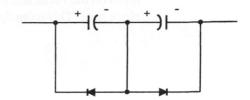

forward-bias the transistor's intrinsic base-emitter zener, you could cause damage. If, instead, the circuit prevents zener breakdown and its clamps prevent excessive reverse V_{BE}, then everything is fine. I probably wasn't clear enough about those points. Note that many discrete transistors are not as fragile—are not going to be damaged or degraded by zener breakdown—as monolithic transistors are. (I'm not sure why.)

Also, you recommend against plugging CMOS components into live (powered up) circuits and working without a ground strap. You say you have seen such procedures directly cause unreliability and failures. OK, you have and other people have, but I haven't. Perhaps RAM chips are more fragile than 74Cxx chips. So, I must retract my cavalier and flippant remarks: In general, you should use ground straps and not plug ICs into live circuits unless you are sure you know what you're doing and are prepared to accept IC failures.

However, when troubleshooting, sometimes you may have to resort to these measures. You should then be aware that they may not necessarily cause harm. But *I*, and the readers, should be aware that sometimes they *may* cause harm, so don't hack around if you don't have to.

As for tantalum capacitors, I've seen very few fail with no provocation. I've used a lot of cheap tantalum capacitors, and they must have been more reliable than I deserved. To ensure that they survive reversal, I suggest the arrangement in Figure 13.3.

Using a current probe to find short circuits is a viable technique, but I don't understand how you can clip a current probe around a board trace. I find that my DC microvolt detector, which appeared in Chapter 2 of the series, will let you track down such DC shorts. If I had to troubleshoot a lot of boards, I'm sure one of your audio-output, milliwatt detectors would be terribly useful.

Although I've never seen a problem with floating TTL gates, you are correct to caution against circuits like the one in Figure 13.2.

RAP

Dear Bob:

As an old hand with 30 years in the business, I have run into some anomalies you did not mention.

1. LS logic is totally unforgiving of negative undershoots at its inputs. The worst chips I have seen are 74LS86s, which hang up for microseconds, totally confusing other circuitry. The second worst is the 74LS75, which can go into either logic state after a negative undershoot but will recover upon the next clock pulse.

2. I have seen circuits with a 7470, -73, -76, -107, -109, -110, or -111 that "remember"

highs on the Preset and Reset lines and toggle in spite of a low-high condition at clock time. This anomaly happens if the clock is left high and is not pulsed.

J. Koontz
Chief Engineer
Computer Automation
Irvine, CA

Dear Bob:

If an engineer wants to see how to properly control EMI at its source, he or she should look at the chassis of any TV and tuner. Chances are, any radiation from internally generated 15-kHz to 950-MHz signals meets FCC Part 15 rules. Obviously, most TV receivers do not depend on their plastic cabinets to contain spurious radiation.

For some time now I have been working with labs that attempt to certify radiating equipment that was not designed using basic radiation-containment methods. These methods date back to the time of tube circuits and include using specialized components such as feedthrough capacitors, ferrite beads, and toroidal coils. These methods also include using very light-gauge, low-cost, tin-plated steel cans to enclose radiating components. Soldering the tabs of these 5-sided cans to the PC board forms a complete 6-sided enclosure around the radiating component. Note that TV receivers' PC boards have considerable ground-plane area.

If you do not design out EMI, someone will have to design in some Band-Aids to fix your bad design.

Here are some points to consider:

1. Incorporate as much ground plane as possible on one side of each PC board that contains digital or analog signals above audio frequencies. Be sure this ground plane has a low-inductance path to the main chassis—even if this path goes through an edge connector.

2. Incorporate a T or LRC filter on all input and output lines. The resistor should always be on the side leading away from the hot circuit. It should be the largest value possible, as great as 1 kΩ. This resistor will damp, or "de-Q," any resonant circuit that the interconnect lines form as well as filter the noise. You must choose the resistor and capacitor carefully so as not to adversely affect the desired signal on the line. The capacitor should be a ceramic disk with a value between 10 pF and 0.01 μF, depending on the signal source. Probe each filtered line to confirm that only required signals are present.

3. Probe each PC board to locate the areas of maximum radiation. Experiment with metal-foil tape to determine where a metal shield will be most effective. Install a temporary shield soldered to the ground plane to verify effectiveness.

4. You should examine purchased items such as disks and power supplies for I/O-line filters and radiation containment.

5. A commercially available filter or a suitable substitute should filter your circuit's input to check that any conducted interference is at least 20 dB below requirements.

6. Monitor your circuit's I/O lines during normal operation and track down any unex-

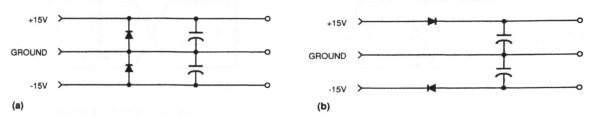

Figure 13.4. Depending on your applications, you may want to connect antireversal diodes across (a) or in a series with (b) a power supply's pins. Or maybe both.

pected signals to their source. Again, any conducted interference should be at least 20 dB below requirements.

7. No matter how you measure the near-field radiation from your design, such radiation should be at least 20 dB below the 3-meter radiation limit. To make these measurements, you'll need a shielded, screened test room.

Thomas L. Fischer
Pacific West Electronics
Costa Mesa, CA

Dear Bob:

You recommend using "antireversal" diodes across a power supply's input to protect circuitry against reversal of the power supply's lead (see Figure 13.4). However, if a power supply does get hooked up backwards, high currents will flow through the diodes, which might degrade or ruin them. Remember, the diodes *are* a part of the circuitry, too. Instead, I recommend connecting the diodes in series with the input pins (Figure 13.4b).

Now the board has protection but with virtually no diode current.

Marvin Smith
Harbor City, CA

Dear Mr. Smith:

You are correct in some cases, and I guess I was delinquent in not mentioning them. For example, if you have a battery, putting the diode in series with the correct path may be appropriate. Then, if reversal happens, the battery won't be crowbarred and you will avoid damage to the battery and it environs. However, if you have a 5-V bus, a diode in series with the supply would both waste a large fraction of the total voltage and possibly spoil the supply's regulation.

Even with a 15-V supply, where the wasted power might be acceptable, the bounce and poor regulation of the supply might hurt the accuracy of the circuits that the supply powers. The diode's impedance may cause poor regulation. So, in cases where a regulator drives the power busses, the shunt antireversal diodes are a good idea. The solution presumes, of course, that the regulators are short-circuit proof.

The worst part of your circuit, Mr. Smith, comes to light when one of the power-supply wires falls off or becomes disconnected. Then, the −14-V bus could get pulled to +5 or +10 V, depending on what loads are between the +14 and −14-V busses. Many linear circuits can get very unhappy if fate pulls their negative-supply pin

Dear Mr. Sturgeon:

Thanks for your tip. Most of the guys in our lab don't use Tempilaq, but it's a good tip. We use thermocouples on the can or diodes in the chip.

RAP

Dear Bob,

So, what do you think about spreadsheets?

Anonymous

Dear Anonymous,

On beds, they are fine, but for linear-circuit design, they can be Bad News. Normally, I hate, despise, and detest them because they give you an answer, but they do not give you a feel for what is important. Also, sometimes spreadsheets lie. When they lie, most people still trust them and never check on them. We have found several cases of a spreadsheet with an error in it, and the error went uncorrected, unsuspected, and even undoubted for a long time. Finally we ran a sanity check and the answer was so silly that we realized nobody had ever checked to see if it even made sense. Like any other form of computer output, you should not trust spreadsheets (and their results) blindly.

RAP

Dear Bob:

You're right. Most people, even technical people who should know better, tend to treat any numeric display or computer readout as if it were engraved in stone, ignoring whatever imperfect mechanism generated digits. They're even in awe of numbers scratched in Jello.

Back in the sixties, there was a story circulating about the Apprentice Engineer who had been experimenting with the plant's new analog computer. He ran up to the Chief Engineer waving a sheaf of printouts. "Look," he said excitedly, "I've come up with a simulation of our power-plant heating and air-conditioning system that will double the plant's efficiency!" The Chief Engineer studied the printout for a few moments. "Yes," he said, "but look here." And he pointed to the flow diagram, "This 17 °F water is going to be awfully hard on the pumps."

Your frustration with the menu burden of these wonderful new instruments is right on target. One thing I detest intensely and that seems to occur with increasing frequency is staring blankly at a screen or cursor, knowing full well that the reason it obstinately refuses to do your bidding is almost certainly a consequence of your imperfect understanding of its menus and possibly also of your failure to read the manual all the way through—which you'll do the next time you have a spare week. Keep up the good work.

Reginald W. Neale
Connoisseur of Solder Globs
Rochester, NY

Dear Mr. Neale,

Thanks for the good words. But, hey, water at 17 °F isn't such a terrible idea if you allow for a little antifreeze. The real problem I foresee is that you can easily get your 17 °F water in the winter, when you don't need it. When you really would like to get some is in the summer, and then it's pretty expensive to chill it down that cold. Maybe all you need is the plumbing equivalent of the solar-powered night-light with the 12,000-mile extension cord.

RAP

Dear Bob:

I'm really confused by your apology on page 34 of the November 23, 1989, issue of EDN. I understand about diode-connected transistors; we use them often. However, we connect the base to the emitter and thereby use the base-collector junction. The breakdown voltage of such a connection is typically about the same as the transistor's maximum V_{CE} rating. If we use your method, the breakdown rating will be only 5 to 7 V typ, according to data sheets. I usually want more.

John Paul Hoffman
Caterpillar Inc.
Peoria, IL

Dear Mr. Hoffman:

You're right, the collector-base junction can handle more voltage than the base-emitter junction, but it's also slower.

RAP

Dear Bob:

A few years ago, when I first moved into a new house and hooked up my stereo, the left channel decided to fry the output transistors. The power amp had heavy feed-

Dear Bob,

You have been writing an excellent series of articles, as you always do. This series reminds me of the old days when I used to be able to fix things. It's necessary to pass this type of wisdom along to each generation to keep common sense in the field of engineering development.

However, as with all us old-timers, you slipped on a few points, to wit: Eyelets are out, especially in multilayer boards. They are still OK for 2-sided experimental boards and single-sided boards where changes in components are desired. In the latter case, an eyelet socket is applied to the finished circuit to hold everything stable. Using an eyelet in a multilayer board will distort the plated barrel, frequently causing separation from the inner layers of copper. This shows up only as a temperature-induced, intermittent problem during testing: It will become a hard failure only when it gets to the customer or is in the field. Good plating shops can create excellent plated-through holes at a reasonable cost today, so give them a try.

The use of lock-washers on printed wiring boards is fraught with danger. We spent a lot of money at one company in trying to find out why our screws kept backing out of their holes, causing their boards to come loose in their mounts. The problems happened during a thermal/vibration test. Running the same boards through the vibration test without the thermal test did not loosen the screws. Thermal cycling was the culprit. It caused the board to expand, and near the glass transition temperature (+125 °C) it caused the board to deform to relieve the stress. When the assembly cooled, there was no incentive for the material to flow back into the gap; thus, the board was loose. A spring washer, be it a star or a Belleville type, will eventually loosen to the point that the desired electrical connection is poor, if not lost altogether. It's better to solder if you can. If a screw is needed, make the pad very large and the washer under the screw wide, in order to spread the load and maximize the joint's life.

Recently I discovered a potential problem with surface-mount capacitors, when soldered onto a PC board. The capacitance can be increased beyond the specifications of the capacitor, if any flux is left hidden under the capacitor. If all the flux is not washed away, it can cause the capacitance to appear out of spec. Washing with solvent and then washing in the dishwasher can solve that problem.

Keep the good practical articles coming, and we will all benefit.

Richard T. Lamoureux
Hawthorne, CA.

RAP's reply:

Well, I believe you are correct, that you are more discerning and have more experience in these areas than I do. Thank you for the tips. You are saying that any bolt on a PC board is likely to loosen unless in an air-conditioned, constant-temperature office with no significant self-heating. H'mmm. Thanks for the tips.

RAP

Dear Bob,

I really enjoyed your recent series on troubleshooting. Although I'm not much of an analog designer, I have gained experience over the years in digital-system design, development, and debugging. Very few people can appreciate the artistic or "seat-of-the-pants" techniques we apply. I recently supervised somebody in the design of a circuit pack. It took him *a few days* to do the actual logic design. It took him several months (on and off) to prepare that design so that somebody could actually build it (a printed circuit board). He was amazed. He never learned any of that in school. He still has to get the circuit pack, get parts (a nontrivial task), and *debug* it. Needless to say, it's quite an educational experience for a novice. Even for an experienced designer like me, it's no trivial task.

This fact is probably responsible for the attitude I have developed over the years: I expect the worst and am surprised when anything works at all. I am rarely disappointed. I design with this philosophy in mind. (I detect this philosophy in Bob's articles, too.)

I run into many people with the opposite frame of mind: They expect the best and are surprised when it doesn't work. I have little patience with these people. They are either geniuses (and I have met two or three of those over the years in this trade), or they've never gone through the design and debugging process. It is truly a humbling process.

One more comment touched on in Bob's article: People are continually amazed when I tell them that I solve most of my problems at my *desk* with an *ohmmeter*. That fact simply points to where half the problems have occurred over the years—in interconnections.

John D. Loop
Research Engineer
BellSouth
Atlanta, GA.

Rap's comment:

I don't find quite that much use for an ohmmeter, but I agree that there are many different ways to find a problem. I think I solve about half of my problems by reading—a schematic, a data book, a customer's request, a spec sheet, a set of test results and test conditions. So I guess you could say that half *my* problems come from bad connections in the way of *communicating*.

RAP

Reference

1. Pease, Robert A., "Protection Circuit Cuts Voltage Loss," *Electronic Design*, June 14, 1990, p. 77.

14. Real Circuits and Real Problems

"Congratulations! You are the proud owner of a brand-new Varoom Automobile. It has been built with the highest old-world craftsmanship and the finest computerized engineering, to assure you of many miles of trouble-free driving. Nothing can go wronX.

"Just in case of some problems, Varoom Motor Co. is pleased to include the following table for troubleshooting:

Troubleshooting Table

Problem: Car will not run.

Indication of Problem:	Solution:
Ash tray is full.	Use Ashtray-Empty computer procedure.
Fuel gauge reads Empty.	Purchase fuel.
Fuel does not reach engine.	Replace Fuel Injection Computer.
Spark does not ignite fuel.	Replace Ignition Computer.
Console display shows "Computer Malfunction."	Have vehicle and checkbook towed to nearest authorized Varoom dealer.

Well, it's nice to know that one out of the five problems can be solved by the car's owner. But personally, I prefer driving a car that can be fixed and troubleshot by ordinary human beings. Go ahead, ask me, what do I drive that fits that description? A 1968 VW Beetle. (My wife has a newer car; she has a 1969 Beetle.) If it doesn't run (which is a fairly rare occasion) I know how to troubleshoot it. Do I look at a table? Yes, but not in a book. I look at a table inside my head. What if I suspect the carburetor or fuel pump? I pour a tablespoon of gas down the carburetor's throat. If it fires and runs and then dies, I know that I can provide gas, but the carburetor can't. Then there are a number of things I can do—such as holding a gallon of gas on top of the car to provide fuel. No pump needed. I haven't had to do that, but I've seen a friend do it.

If I suspect the spark, I always have enough spare parts that I can check the spark, check the timing, swap in spare points, rotate the distributor, whatever is needed. I keep spare parts in my car, to help a friend when I spot a VW broken down along the highway. I even keep a spare envelope (SASE) in the car so if I try to help a guy and we can't fix it, and we can't figure out why, I give him the envelope and he can mail it back to me to tell me what the problem turned out to be.

Last year, I got a letter back from a guy who "broke down" along Bayshore with a newly-purchased 1970 VW Bus. At the time, we couldn't figure out what was wrong with it. In the letter, he explained that the gas gauge always read 3/4 full because it was broken, which was why he ran it out of gas and didn't know it. Unfortunately, the guy he bought it from wasn't helpful enough to warn him.

So, if I keep on driving this basic car forever, I will know everything I need to

172

know about it. (Go ahead. Say, "OK, Pease, you can't keep on driving a 1968 VW forever." You can say that, but you are wrong. I can buy enough 1968 VWs to last me for another 50 years. There are a *lot* of 1968 VWs in very good shape, here in California. . . .)

Back to Electronic Circuits . . .

Just as many cars are designed to be repaired by swapping out a large modular segment and swapping in a replacement, so many electrical circuits and systems are designed with swap-in cards, which are "non-field-repairable." Even circuits which are pretty easy to repair are said, by habit, to be nonrepairable. In fact, the advent of "throw-away" modules has been debated. Personally I don't approve of it, not a darned bit. A couple weeks ago, this little Compaq portable computer quit—the one I'm using for word-processing. When I tried to read its technical literature and find out its advice on how to repair the computer, it told me to use various software inquiries to find out the problem. What a completely useless notion—the CRT and all other functions were dead, so I couldn't *possibly* use its diagnostic software. *Fortunately* I have a couple technicians who cannot imagine the meaning of "cannot be repaired." Paul climbed in and found a shorted rectifier, replaced it, and I was back on the air in a couple days. If I'd had to take it to a repair shop, I hate to think of the $ and days just to get a $2.00 rectifier replaced. I'm sure the power-supply card would cost $90, not to mention the labor.

When I was wandering through Kathmandu last year, I saw workers repairing things that would not be worth the effort in the USA. But Nepalis do not have enough money to join in the "throw-away" society, so they make the effort to repair things. Cars, tires, stoves, tools—any equipment that could possibly be repaired, usually is. (And if it can't be repaired, it often gets recycled.) I support that approach, and I myself am usually willing to put in a lot more hours of effort to troubleshoot something, than the cost of replacing it would justify. Why? Because, sometimes I learn something.

Once I had an old 1970 VW which I retired because it had 249,850 miles on it and because it was leaking oil badly, from a cracked block, or so I thought. When I actually started to dismantle the engine, I found, not a cracked block, but that the bolts that fasten down the oil cooler had come completely loose. And why had they come loose? Because there were no lock-washers on the nuts. So in the future I made sure that anybody working on my engine would use lock-washers, to minimize the chance of nuts coming loose. It was an educational experience, and well worth the effort.

So, let's presume that we may actually do some troubleshooting and repair, rather than just chuck the circuit in a wastebasket. I was talking with a guy the other day, and he said, "Bob, be sure to spell out the difference between Lab Troubleshooting and Production Troubleshooting." I don't think I know what that difference is. In either case, it can be very important, and a small amount of time and money can have great returns. Of course, on other occasions, you can put in a lot of hours and get virtually nowhere. . . .

As with any other system, troubleshooting is an art which can be developed with practice. You have to learn the failure modes, the patterns of abuse, the procedures for replacing bad parts, the documentation, and all the other things we have discussed. But how about a Modus Operandi? Let's look at the following table for a simple op-amp inverter:

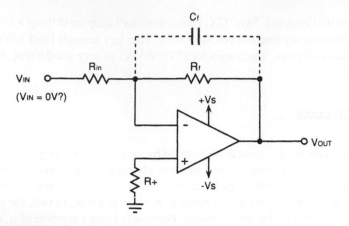

Figure 14.1. Basic Op-Amp Inverter.

Table 14.1 Troubleshooting Inverting Operational Amplifiers

Indication of Trouble	Possible Cause	Solution
Output voltage offset is excessive (when input is zero).	Feedback R too high?	Use lower R_f, or better op amp with lower I_b.
	Oscillation causes offset?	Use scope, check for oscillation.
	Sneak path for leakage into input?	Check for dirty, leaky PC board or connectors.
	R+ is wrong value?	Make R+ = $R_{in} \parallel R_f$.
	Op amp's V_{OS} too big?	V_{out} should be < V_{OS} x $[(R_f/R_{in}) + 1]$. Use a V_{OS} trimpot *or* a better amplifier for low V_{OS}.
	Amplifier out of spec?	If all other causes are negative, remove op amp and test it.
Output pegged near one power supply rail.	Other power supply may be missing?	Check voltage on each pin of the part, *not* just the PC board.
	Output is shorted?	See if amplifier is hot. Check continuity.
	Bad amplifier?	Pull out unit, test it.
Output is oscillating?	Input oscillating?	Check input.
	Power supply oscillating?	Check each supply.
	P.S. bypass caps missing or inadequate?	Try more caps, closer to unit, or bigger or better ones.
	Cap. load too heavy?	Look for cables; measure the C load.
	No feedback cap?	See text; try different. values of C_f.
	Oscillation in air?	Turn off power and watch.
	Comp cap too small? (LM301A or similar)	Try adding more capacitance.

(Continued)

	Output oscillation is intermittent?	Check to see if output is ringing (see Pease's Principle).
Output distorts?	Load too heavy?	Check resistive *and* reactive load.
	Input is distorted?	Check the input.
	Slew rate distortion?	Test with a lower input frequency or size.
Bad gain?	Resistors have bad tolerance or wrong value?	Check resistor markings and tolerances.
	Oscillations at various levels?	Check for oscillations across working range.
In general:	Amplifier is suspected to be bad?	Swap in a known good amplifier.
	Swapped amplifier is "bad," too?	Swap "bad" amplifier into a good circuit.
"No output?" (Output is zero volts).	Output shorted to ground?	Amplifier gets hot. Turn off power, measure ohms.
etc., etc.	Amplifier has low V_{OS} (a very good op amp)?	Put in a signal thru R, see if output moves.

Now, what is the best thing about this table? That it will solve all your op-amp problems? Hell, *no*!! You can surely bump into circuits and problems that I have never seen, that I have never even envisioned—circuits that need more help than this table will give.

Well, is it because it gives you some general approaches that can be used for any circuit? That is a good idea, and this is definitely of some value, but *that* is not the most valuable thing.

Okay, what *is* the most valuable thing about this table? The most valuable thing is that *you* can make up your own troubleshooting tables. You don't have to be perfect, or brilliant, or unerring. You don't have to keep perfect notes. You don't have to make a plan of action and follow it exactly, one after another sequentially. You don't even have to write your plans down, although that *is* usually a good idea. You don't have to do any one thing, except to *think* occasionally. If you do some thinking, in a skeptical way, you can guess solutions and tests and answers that would take me forever. You have your own systems with which you are familiar, and your equipment, and your friends. Together, you can solve problems that nobody else can. So, I guess I'll admit that some confidence would be a useful tool for you. And if there are specialized techniques that you know, well, good for you. I never told you that I know everything. But I bet some of the techniques in this book will be useful.

I will throw in a couple more scenarios for other basic circuits. They may not solve every problem, but they will indicate the breadth and depth of thinking that may be needed to solve tough circuit problems.

Examples:

- Single-transistor amplifier
- Negative regulator (with LM337)
- Absolute-value circuit
- Switching regulator using LM3524
- Positive regulator (with LM317)
- 723-type regulator
- Instrumentation Amplifier
- Switching regulator using LM2575.

Figure 14.2. Basic Single-Transistor Amplifier.

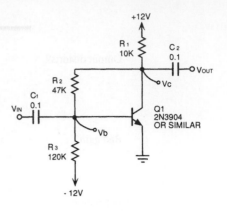

Table 14.2 Troubleshooting Single-Transistor Amplifiers

Indication of Trouble	Possible Cause	Solution
Output at wrong DC level		
Collector at +12 V?	R2 broken or missing?	Check ohms and volts. Touch in 47 k across R2.
	R3 shorted?	Look for shorted foil.
	R1 shorted?	Look for shorted foil.
Collector at +10 V?	Q1's base shorted?	Measure base voltage.
	Q1's collector open?	Check c-b diode.
Collector at +0.7 V?	R3 broken or missing or open?	Check resistor. Touch in 120 k across R3.
	R1 or R2 has wrong value?	Check resistors.
	Collector-base short?	Look for shorted foil or transistor.
Collector at 0 V?	Short from collector to ground?	Look for shorted foil or shorted transistor.
Base at wrong voltage		
Base at +3 V?	Solder connection missing?	Make sure base and emitter are actually connected.
	Base-emitter junction probably blown?	Replace Q1.
	PNP transistor?	Double-check.
Base at −3 V?	Wrong resistor values?	Check resistor values.
	Input signal much too big?	Check with scope.
No gain or bad gain	Capacitor missing or too small?	Add a good capacitor across C1 or C2.
	Bad DC bias?	Check for DC levels, as above.
	Q1 installed backwards?	Check.
Oscillation.	General:	Study frequency of osc'n.
	Power supply oscillating?	Check power supply; add more bypasses.
	Load is oscillating?	Short Q1 base to ground, look at collector?
	Load causes Q1 to oscillate?	Remove load, study load.
Low gain (non-inverting)	Transistor broken?	Check as above.
	PNP transistor?	Check.
etc., etc., etc.		

Comments on Troubleshooting Table for Single-transistor Amplifiers

This circuit has many things in common with many transistor circuits. It is a good circuit to practice and sharpen up your troubleshooting skills on. A DVM is not a useless tool, but a scope does a lot of things better.

Now let's move on to the Positive (Adjustable) Regulator.

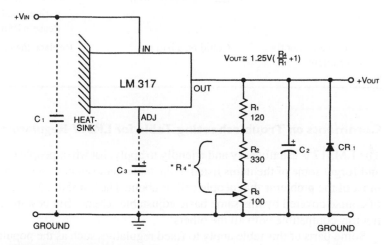

Figure 14.3. Positive Adjustable Regulator.

Table 14.3 Troubleshooting Positive Adjustable Regulators

Indication of Trouble	Possible Cause	Solution
Output V much too low.	Input much too low?	Check input with scope.
	R2 or R3 shorted?	Check resistors.
	C3 shorted or reversed?	Check C3 or remove.
	Load is short-circuit or too heavy?	Unit gets hot. Try disconnecting load.
	CR1 is shorted, or backwards?	Check diode.
	C2 shorted or reversed?	Check C2 or remove.
	Leakage at ADJ pin?	Check for dirty board.
	Output −0.8 V?	Output shorted to − supply.
Output V much too high.	Pot wiper open?	Check voltages on pot.
	R1 too low?	Check value.
	Input V much too high?	Check with scope.
	Leakage at ADJ pin?	Check for dirty board.
Output oscillates.	General:	Note the frequency.
	Check for C1?	C1 should be 0.1 μF or more, see data sheet.
	Load bounces too much?	Add a lot of C2 and see what helps.
Output too noisy.	Input too noisy?	Study noise on input. Increase C1, C2, or C3.
	Load too noisy or jumpy?	Increase C2 and see what happens; also C1 or C3.

Output drifts, falls out of regulation.	Unit much too hot?	Increase the heat sink. Study the dissipation= $(I_{load}) \times (V_{in}-V_{out})$.
Output shifts badly vs. I_{load}.	R1 connected too close to load? (Bad Kelvin connections)	Connect R1 to LM317's output pin directly, (use different wire than for the load).
	Load causes oscill'n?	Monitor with scope at various loads; try Pease's Principle.
Items above do not fix the problem. etc., etc.	Could be a bad part?	Replace the regulator.

Comments on Troubleshooting Table for LM317 Regulators

The LM317 is usually easy and friendly to apply, but when people get absent-minded and forget some of the items listed in the table, they can have problems. This covers most of the problems that people call us about. The LM350, LM338, and LM396 are of course covered by this same basic adjustable scheme. But beware, as the LM396 is not pin-compatible with all the others!

Some parts of this table apply to fixed regulators, such as the popular LM340s and LM7800s which are available in 5-V, 12-V, and 15-V versions.

If you use a circuit like this a lot, you ought to have a little breadboard with a socket so you can check the IC to see if the problem follows the IC, or stays with the circuit. When you do that, remember that the load regulation will probably be mediocre unless you have a good Kelvin socket; and the part will probably get hot quickly if it has no heat sink. Now let's move on to the Negative Regulator, which is substantially the same as the Positive Regulator, so we will put into the table only the items that are different.

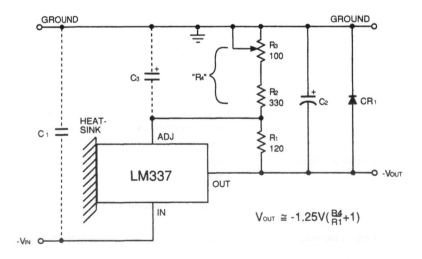

Figure 14.4. Negative Adjustable Regulator.

Table 14.4 Troubleshooting Negative (Adjustable) Regulators

Indication of Trouble	Possible Cause	Solution
Bad DC errors.	Refer to table for positive regulators, above.	
Oscillation.	Refer to table for positive regulators, ALSO:	
	Improper C2 capacitor?	Refer to manufacturer's data sheet: Avoid ceramic or film caps. Use tantalums or big electrolytics.
	Too many ceramic discs connected on output bus?	Swamp them with over-compensation, with tens of μF tantalum, or a few hundred μF aluminum electrolytic.

Comments on Table for Negative (Adjustable) Regulators

These regulators have just about all the virtues and the freedom from problems as the positive regulators, but they are quite critical about having a good capacitor from the output to ground, as noted above.

The fixed negative regulators (LM320 family, LM7900 family) are likewise quite demanding that a decent capacitor be used for the output's damping. This is inherent because the negative regulators all have collector-loaded outputs, and you need a good capacitor to roll off the extra gain.

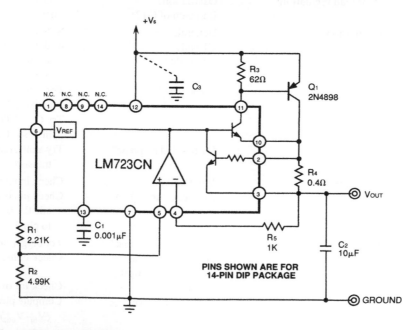

Figure 14.5. LM723-Type Regulator.

Table 14.5 Troubleshooting LM723-type Regulators

Indication of Trouble	Possible Cause	Solution
Output voltage has small error	Resistors incorrect?	Check R1, R2.
	Oscillation?	Check for oscillation.
	Reference bad?	Read V at pin 6.
	Amplifier bad?	Check V at pins 4, 5.
Output voltage much too low	Resistors incorrect?	Check R1, R2. Check V at pins 6, 5, 4, 3
	Reference bad?	Check V at pin 6
	Comp cap shorted?	Check V, ohms at pin 13.
	Q1 dead?	Check V at pins 12, 11. Pull out Q1, replace.
	R4 broken?	Measure R4 ohms.
	Load too heavy?	Measure V across R4. Remove load.
	Input much too low?	Check V_{in} with scope.
	PC foil shorted?	Review all data above.
	PC foil open?	Review all data above.
	LM723 busted?	Review all data above. Test, replace IC.
Output voltage much too high.	Same as above.	Same as above.
	Q1 shorted out?	Check voltage from pin 11 to 12. If that's OFF, transistor should be OFF.
	Input V too high?	Check V_{in} with scope.
Output cannot drive rated load.	R4 wrong value?	Measure R4.
	Q1 broken?	Check, replace Q1.
	Something bad?	Review all data above.
Short-circuit current too big.	Current limiter busted?	Check voltages at pins 2, 3. Check R4. Replace 723.
Poor load regulation.	Oscillation?	Check for oscillation.
	Bad gain of Q1?	Check and replace Q1.
Oscillation.	General:	Note the frequency.
	C1 bad?	Add .001 across C1.
	C2 needed?	Add various capacitors across C2, 10 to 500 μF.
	C3 needed?	Try various caps at C3, 0.1 or 100 μF or both.
		Try C2 AND C3 and also increase C1.
	Q1 has bad response?	Try a different type of transistor.
Output noisy.	Input noisy?	Check input with scope.
	Reference noisy?	Check pin 6 with scope, add 0.1 or 1 μF mylar to pin 5.
Q1 dies at full load.	Inadequate heat sink?	Try bigger heat sink.
	Heat sink bolts too loose or too tight?	Check bolts.
	Oscillation?	Check for oscillations.
	Power excessive?	Compute power, IL × (V_{in}–V_{out}).

etc., etc., etc., etc.

Comments on Troubleshooting Table for LM723 Regulators

As you can see, there are, potentially, a lot of things to worry about. The LM723 is rarely used these days unless a specific feature requires it. I don't want to scare you, but anybody who troubleshoots a number of old-style regulators has to understand the circuit thoroughly, so he can tell immediately when he looks at a couple voltages, whether he is on the right track. He has to have the concepts behind the chart built right into his head, or it would take forever to troubleshoot a basketful of bad modules. (Okay, *he* or *she*. . . .)

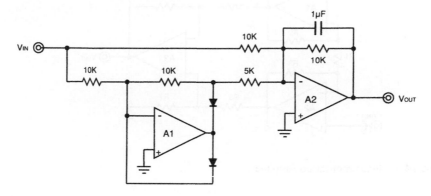

Figure 14.6. Full-Wave Rectifier.

Table 14.6 Troubleshooting Full-wave Rectifiers

Indication of Trouble	Possible Cause	Solution
Input amplifier runs badly.	Anything?	Apply V_{in} = +0.1 to +10 VDC and troubleshoot it per the procedure above, for inverting op amps.
Output amplifier runs badly.	Anything?	Apply V_{in} = –0.1 to –10 VDC and troubleshoot it per the procedure above, for inverting op amps.
	Bad diodes?	Check for reversed or shorted diodes. Watch A1 output with scope.
Bad AC response.	Diodes too slow?	Check out diodes in little sockets; compare to known good diodes.
	Amplifier too slow?	Check amplifier with large and small AC signals, fast and slow.
Bad DC errors hot.	Diodes too leaky?	Compare leakage to known good diodes.

Comments on Table for Full-wave Rectifiers

As with other complicated circuits, if you have to keep a circuit even as complex as this in production, you should have a breadboard all built up, with sockets for any critical components, to make it easy to evaluate them at a minute's notice. Otherwise you may just try to duck the problem, and that would be wrong.

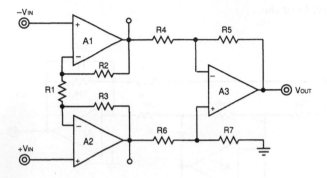

Figure 14.7. Instrumentation Amplifier.

Table 14.7 Troubleshooting Instrumentation Amplifiers

Indication of Trouble	Possible Cause	Solution
Input stage works badly.	Anything?	Ground one input and put a signal in the other input; troubleshoot as inverting amplifier. Then swap inputs.
	Bad output stage?	Ground one input of output stage and put a signal in the other; troubleshoot as above.
Bad DC errors.	Anything?	Ground both inputs; read all voltages with DVM; remove suspected bad amplifier and test.
Bad CMRR.	Input stage?	Tie both inputs together and drive + and –. Read all voltages. Check input op amps' CMRR.
	Output stage?	Check resistors' match and trim range. Check output op amp.

Comments on Table for Instrumentation Amplifiers

As above, this circuit should be set up with sockets for ease of evaluation. This circuit does offer a little more interaction, but it is not really too difficult when you figure out what has to be going on.

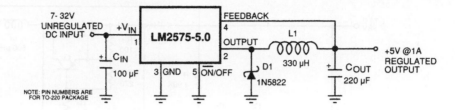

Figure 14.8. LM2575 Switching Regulator.

Table 14.8 Troubleshooting LM2575 Switching Regulators

Indication of Trouble	Possible Cause	Solution
Output V much too low.	Input V too low?	Check V_{in} with scope.
	Output shorted or overloaded?	Remove load. Check ohms, output to ground.
	ON/OFF pin not connected?	Check voltage at pin 5.
	Bad rectifier?	Check volts, ohms on D1.
	Bad IC?	Check, replace, swap.
	Ambient temp too hot?	Check ambient, cool it.
Output V much too high.	Feedback path bad?	Check V at pin 4.
	Bad inductor?	Check, swap.
Inductor overheats?	Frequency too high?	Check frequency at pin 2.
	Bad inductor?	Check L, check its lossiness vs. a known good inductor.
	Shorted rectifier?	Check rectifier.
IC overheats.	Frequency too high?	Check frequency at pin 2.
	Output shorted?	Check V_{out}, I_{load}, I_{supply}.
Rectifier overheats?	Frequency too high?	Check frequency at pin 2.
	Diode too slow?	Compare to good diode.
Bad ripple.	Frequency bad?	Check frequency at pin 2.
	Bad capacitor?	Check capacitor for Rs.
	Wrong mode?	Check waveforms.
etc., etc.		

Comments on Table for LM2575 Switching Regulator

Even though this is a very simple-looking circuit, with only a few parts, you should keep a breadboard (with sockets) around because the inductor is such a tricky component.

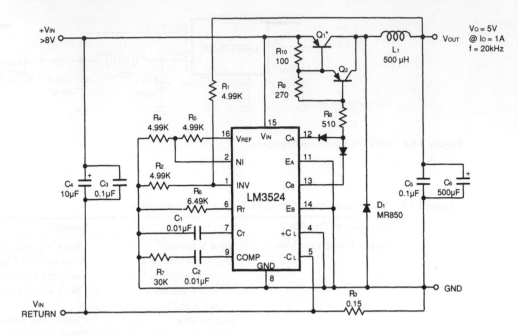

*MOUNTED ON STAVER HEATSINK No. V5-1.
Q1 = BD344
Q2 = 2N5023
L1 = >40 TURNS No. 22 WIRE ON FERROXCUBE No. K300502 TOROID CORE.

Figure 14.9. LM3524 Switching Regulator.

Table 14.9   Troubleshooting LM3524 Switching Regulators, SIMPLIFIED

Indication of Trouble	Possible Cause	Solution
Something overheats.	Frequency too high?	Measure F at pin 7.
	Bad D1 rectifier?	Check D1.
	Bad transistors?	Check transistors.
	Bad capacitor?	Check capacitor.
	Bad LM3524?	Check LM3524.
Clock frequency bad.	Bad capacitor?	Check Ct at pin 7.
	Bad resistor?	Check Rt at pin 6.
Bad loop stability.	Bad R-C damper?	Check R-C at pin 9.
	Bad C6?	Swap a good cap.
	Bad inductor?	Check L1.
Bad current limit.	Bad R_{sense}?	Check R_{sense} at pins 4, 5.
Output voltage tolerance is bad.	Bad resistors?	Check R1, 2, 4, 5.
	Bad reference?	Check V at pins 16, 2, 1.
Transistor overheats.	Resistors bad?	Check R8, 9, 10.
	Bad inductor?	Check L1.
Nothing works.	Solder shorts?	Look for solder shorts.
	Something bad.	Check everything.
Can't drive rated load.	Bad transistor?	Check Q1, Q2. Swap in good transistors.
	Bad input voltage?	Check V_{in} with scope.
	Bad 3524?	Study all data, swap in a known good part.

etc., etc., etc., etc., etc., etc. . .

Comments on Table for LM3524 Switching Regulator:

Obviously, if a circuit of this complexity gives problems, you will need to be almost as wise as the person who designed it. (Or maybe even wiser?) And, you will need to memorize what all the correct waveforms ought to look like, so you can detect errors and deviations. You will need a compact breadboard so you can evaluate the critical components in sockets. There may be, in any given batch of badly running parts, several different modes of unhappiness. If you have to troubleshoot a lot of these, you will be a good troubleshooter when you are done.

NOTE: This circuit is based on Figure 15 from the 1989 National Semiconductor LM3524 data sheet, which *did not work* because the diodes to pins 12 and 13 were drawn backwards . . . sorry about that. I noticed this problem even before I started to build this. Then I did build this. It ran. Anyhow, now you know why the LM2575 is called a "Simple Switcher"— because of comparison to old circuits like this.

Final Floobydust

I almost forgot this, but when you make up these little breadboards for evaluating circuits and their components, you may need some tiny connectors for diodes or transistors or small capacitors. You cannot use those nylon breadboarding panels, as mentioned in Chapter 13, because the capacitances and inductances will be hopelessly bad. You just about have to use the same basic PC board that is used for the real circuit, and then install tiny component jacks that accept 0.018-inch leads, such as Amp[1] type 50462, or Interconnection Products[2] type 450-2598-01-03-00.

If it's inconvenient to go out and shop for these component jacks, you can "roll your own": get some of those strips of sockets, the ones that are 25 in a line, such as Digikey's Catalog No. A208, A209, or similar.[3] If strays are not important, just snip off a group of jacks—as many sockets as you need in a row. But if you need low capacitance and very low leakage, just use your diagonal nippers to snip away the plastic, and use the little jacks one by one. You will want to avoid beating these up, as they are a little delicate, but they are excellent sockets for diodes, Rs, Cs, Qs, and other small components with thin leads. If you need a little component jack for 0.040-inch diameter pins, the Interconnection Products type 450-3729-01-03-00 is good for that, and a similar part from Amp is 645-508-1. The main point is, you want to affect the stray capacitances and inductances of the real circuit as little as possible, so when you stuff it full of good parts, it works OK. Then when you swap in a bad part, it should be obvious what's to blame.

Here's a *final* list of some of the most common problems in circuits:

- Swapped resistors—installed in the wrong place

- Resistors of the wrong value—wrong code

- Diodes installed backwards

- Electrolytic capacitors installed backwards

- Broken wires

- Links installed in the wrong place, or, missing

- Flakey connector

1. AMP, P. O. Box 3608, Harrisburg, PA. 17105. (717) 564-0100.
2. Interconnection Products Inc., 2601 South Garnsey, Santa Ana, CA 92797 (714) 540-9256.
3. Digikey, P. O. Box 677, Thief River Falls, Minnesota 56701-0677, (800) 344-4539.

- Solder shorts

- Unsoldered joints, cold-soldered joints

- IC plugged in with the wrong pins

- Transistors backwards

- Faulty meter or other tester

   If you can check for these problems, you have a good chance of catching 50% of the problems . . . and then the really tough ones will be left over for you to solve. Best of luck!

Aptitude and Attitude

If you have to treat your troubleshooting work as a chore, as a necessary but boring task, well, I can't tell you how to do your work; but I have always found that if you can treat any such work as a *game*, then you have a chance to bring more innovative approaches to the job, and chances of doing it better and having fun. I mean, in my job, if I get the job done, but I am not having any fun, then something is wrong. Hell, sitting here typing this at 12:05 AM is fun! If it weren't, I would stop and go to bed. But, hey, fun or not, this is the *end*, so, *good-night*!

RAP

Appendix A

Digital ICs with Nonstandard Pinouts

I printed this little list of ICs that have nonstandard pinouts in *EDN Magazine*, back in the fall of 1990, saying, "Tell me if this list is incorrect or incomplete," but I have not heard any complaints.

Example: The DM7486 and DM74S86 and DM74LS86 and MM74C386 all have the same pinout. But the DM74L86 and MM74C86 have a *different* pinout. When the L86 came out, it was not pin-compatible with the 7486, and of course the C86 came out to be compatible with the L86. Several of these parts are rare and obsolescent and sole-sourced, and you will probably never see one, but they are listed here for completeness.

Here is the complete list:

- 74H01
- 74L51, LS51
- 74H53
- 74L54, LS54, H54
- 74H55
- 74L71
- 74L78, LS78A
- 74L85, C85
- 74L86, C86
- 74L95, C95

Readers might also want to be aware that the DM74107 (for example) is pulse-triggered, but the 74LS107 is edge-triggered, so in some cases they might not be exactly interchangeable. There are other similar cases, which can be gleaned from the fine print of the data sheets of digital ICs. Refer also to Chapter 10, pages 123–124, for a discussion of other incompatibilities of digital ICs.

Appendix B

Operational Amplifiers with Nonstandard Pinouts

The other day I was writing to an old-timer and explaining that the new IC op amps are not only as easy to apply as the old vacuum-tube ones, but in many respects easier. I began to explain that the LM607 is easy to use and to trim for offset voltage, and so is the LF355, and—woops! Yes, each one is easy to trim, but they have different trim schemes. The LF356 needs a trim pot from pins 1 and 5 to $+V_S$, whereas the LM607 needs a pot from pins 1 and 8. Furthermore I recall some people griping that NSC's LF351 has trim pins for which you turn the trim pot *one* way to increase the V_{OS} and the LF411 has trim pins for which you turn it the *other* way. Then one time a few years back, we started shipping LF411s marked as LF351, as an "improvement." Oh, yes, many customer were probably pleased by the "improvements," but some customers were peeved because we had reversed the gain of their trim circuit. So why do I spend my time griping about MM74C86s that have nonstandard pinouts, and ignoring the problems with linear ICs?!?! So, here you are.

First item, these lists apply only to single operational amplifiers in 8-lead packages, metal or plastic. Next, the pins 2, 3 and 6 are *always* the inputs and output. Next, pins 4 and 7 are *always* the − and + supplies. Next, the 8th pin—the one that's left after the two trim pins are assigned—is almost never standardized, and I am not even going to mention it here. In some cases it's an over-comp pin, and other times it has other functions. No comment.

Nonstandard Pinouts:

- LM709
- LM107, LM307
- LM118, LM318
- LM10

- LM101, LM301
- LM108, LM308
- LM144, LM344
- LH0024

(These are all mongrel and nonstandard compared to *anything* else. Refer to the individual data sheets.)

TYPE I—10-kΩ pot to −V_S from pins 1 and 5.

- LM741
- LM143 (100 kΩ)
- LM776 (100 kΩ)
- LM4250 (100 kΩ)
- LF351
- LF411
- LF441

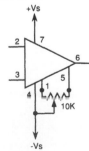

Figure B.I. Type I: Pins 1 and 5 Trim to −V_S.

- LF451 (in SO-8 package)
- LF13741
- LH0022
- LH0042
- LH0052

TYPE II—25-kΩ pot to +V$_S$ from pins 1 and 5.

- LF156, LF356
- LF155, LF355
- LF157, LF357
- LF400   (10 kΩ)

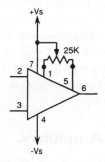

Figure B.2.   Type II: Pins 1 and 5 Trim to +V$_S$.

TYPE III—10-kΩ pot to +V$_S$ from pins 1 and 8

- LM11 (100 kΩ)
- LM112 (100 kΩ)
- LM607
- LM627
- LM637
- LM725
- OP-07 (20 kΩ)

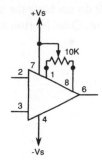

Figure B.3.   Type III: Pins 1 and 8 Trim to +V$_S$.

TYPE IV—100-kΩ pot to –V$_S$ from pins 1 and 8

- LM6161, LM6361
- LM6162, LM6362
- LM6164, LM6364
- LM6165, LM6365

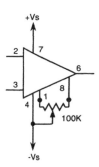

Figure B.4.   Type IV: Pins 1 and 8 Trim to –V$_S$.

Now, it is true that I have only listed NSC parts in here. Maybe someday we will list all the parts in the world, but I haven't the time to do that now. I'll leave enough space for you to put your notes alongside these for the sake of completeness. (Note, in no case can I vouch for which direction you turn the pot to increase V_{OS}. That's for you to determine.)

Notes On Dual Amplifiers.

All DUAL operational amplifiers in 8-lead TO-99 packages (often called "8-lead TO-5") *or* 8-pin mini-dip *or* SO-8—all have the same pinout, as far as I have ever seen. Examples are: LM158, LM833, LM1558, LM6218, LF412, LF442, LF453, LMC662, LPC662, etc., etc. There are no exceptions that I know of—tell me if I am wrong—in this field. (The old dual LM747 in its 10-pin package, or the dual LM709 in 14-pin packages or any dual in any package with more than 8 pins—they do not count.)

Notes on Quad Amplifiers.

All quad operational amplifiers in 14-pin DIP packages *all* have the same pinout, EXCEPT FOR THE LM4136!!

The standard ones include:

LM124/LM324, LP324, LM349, LM837, LF347, LF444, LMC660, LPC660, etc. There are very few exceptions in this field. (Note—Quad comparators such as LM339 do not have the same pins as quad op amps. Even the power supplies are different. Quad Norton amplifiers such as LM2900 are also nonstandard.)

Appendix C

Understanding and Reducing Noise Voltage on 3-Terminal Voltage Regulators

**Erroll H. Dietz, Senior Technician,
National Semiconductor Corporation**

The usual approach to reducing noise on 3-terminal voltage regulators has been to simply place capacitors on the output and on the adjust pin for adjustable regulators. As it turns out, the addition of output capacitance on most voltage regulators may reduce the noise over a broad frequency range but may increase noise within a narrow frequency range. Since the output impedance of most 3-terminal regulators is inductively reactive over a certain frequency range, one can surmise that adding output capacitance to improve noise performance and transient response can also have other effects. The examples given in this appendix will use the LM317 adjustable voltage regulator, but this information can be scaled and then applied to all other kinds of 3-terminal voltage regulators.

As shown in Figure C.1, the output impedance of the LM317 over the 1 kHz to 1 MHz frequency range is inductive. This has nothing to do with long wires, but is simply another way of looking at the fact that the gain of any operational amplifier or

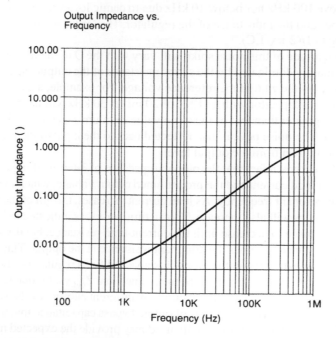

Output Impedance vs. Frequency

Figure C.1. LM317's output impedance vs. frequency at I_L = 500 mA.

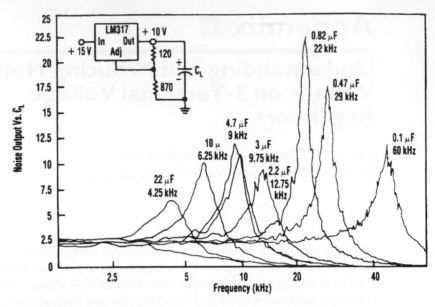

Figure C.2. Typical noise peaks produced by an LM317 and various capacitive loads.

regulator is designed to roll off at 6 dB per octave. This condition is usually not of much concern to the average user of IC regulator circuits. However, this inductive output impedance coupled with an output capacitor to ground can produce a noise peak within a narrow frequency range. This noise peak coincides with the resonant frequency of the inductive output impedance of the regulator and the load capacitance on the output. Figure C.2 shows typical noise peaks produced by an LM317 and various capacitive loads. The frequency range of the noise spike does not extend much above 100 kHz nor below 10 kHz due to ohmic losses in the added output capacitance and the inductance of the regulator. The frequency is predictable according to $1/(2\pi\sqrt{LC})$.

The magnitude of this noise spike will vary with the Q of the resonant circuit, which is mainly dominated by the series resistance of the output capacitor and is proportional to the gain of the reference voltage. For example, a good 1 μF polypropylene capacitor with an ESR of 20 mΩ at 30 kHz will have a noise peak three times greater than that of the same value tantalum capacitor which has an ESR of 1 or 2 Ω. The noise peak is also reflected back to the input of the regulator and is about 20 dB down from the output level.

It is a little-known fact that the output impedance of 3-terminal regulators can vary greatly with load current and the programmed output voltage, which in turn varies the noise peak resonant frequency. As load current increases, the gm of the regulator's output transistor will also increase. This in turn causes, as indicated in Figure C.3, the output inductance to decrease until the current-limit resistance, bond-wire resistance, and lead resistance becomes the dominant resistance at the output. This is true for positive and negative regulators, for adjustable and fixed regulators, and for large and small regulators. In the past it has been assumed that Z_{out} vs. Frequency was a fixed curve, but really there is a family of curves at different current levels (see Figure C.4).

In conclusion, the typical values of output bypass capacitance that users of 3-terminal regulators have traditionally used may provide the expected noise reduction at some frequencies, but not at every frequency. In most cases, a few microvolts of power-supply noise peaking at 5 or 10 kHz will not cause any problems. However, if

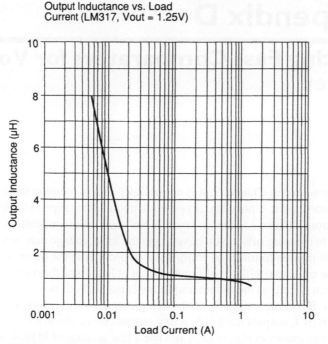

Figure C.3. LM317's output inductance vs. load.

the application circuit is extremely sensitive to excess noise on the supply at one particular frequency, then the user can easily select an appropriate output bypass capacitor and engineer the regulator's circuit so that the noise peaking will fall outside the critical range. Capacitors that fall into the range of 0.1 to 20 μF should be avoided in low-noise applications, especially those with low ESR. The most effective noise reduction can be realized when an electrolytic capacitance of 50 μF or greater is placed at the output, and at least 1 μF at the ADJUST pin, for adjustable regulators. The user should also be aware that changing the load current or output voltage can change the output inductance, so the circuit must be evaluated over the complete range of load currents and output voltages over which the regulator will be operated.

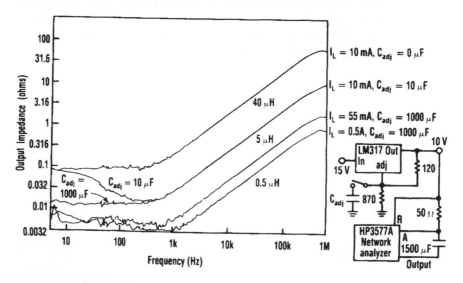

Figure C.4. Output impedance vs. frequency at different current levels.

Appendix D

Testing Fast Comparators for Voltage Offset

As mentioned in Chapter 9, it is not trivially easy to measure the V_{offset} of fast comparators, but it is possible, if you think about all the aspects of the problem. Fast comparators all have a tendency to oscillate when their input voltage is nearly zero. And yet to measure offset, you have to get the input voltage near zero. The solution to this dilemma is to force the comparator to oscillate at the frequency *you* define. The basic op-amp oscillator of Figure D.1 is able to force the oscillation, but it is not a precision circuit, as the output amplitude is not well defined.

In fact, fast comparators do not have big output swings or symmetrical swings; the ones with ECL outputs have only a tiny output. So we add in some gain from an LM311, (as shown in Figure D.2) and use a few sections of MM74C04 to provide a symmetrical output. This circuit puts ± 10 mV at the + input of the DUT, and the waveform at the – input of the comparator is thus forced to ramp back and forth between $(+10\,\text{mV} + V_{OS} - V_{noise}$ and $(-10\,\text{mV} + V_{OS} + V_{noise})$. The average value of the – input's voltage is thus equal to the V_{OS}, as required.

Now, this circuit's offset would not be true if the LM311 had a bad delay in one direction, and a worse delay in the other direction, and that would be the case *if* we did not include some fast AC-coupled hysteresis, per the 4.7 kΩ/100 pF network. This forces the DUT to turn around and integrate back the other way as soon as its input hits the threshold and its fast output responds, without waiting for the slower response of the LM311. As mentioned in Chapter 9, this AC-coupled hysteresis decays and has no effect on the accuracy of the oscillator.

The circuit of Figure D.3 is very similar but is adapted for comparators with an ECL output such as μA6685. The LM311's threshold is changed, and the amount of AC hysteresis is maintained by changing the impedance. Oscillation occurs at 0.4 MHz. No spurious oscillations have been observed, although as with any fast circuit, a thoughtful layout is mandatory.

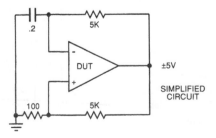

Figure D.1. The basic concept of a self-oscillating test circuit is familiar.

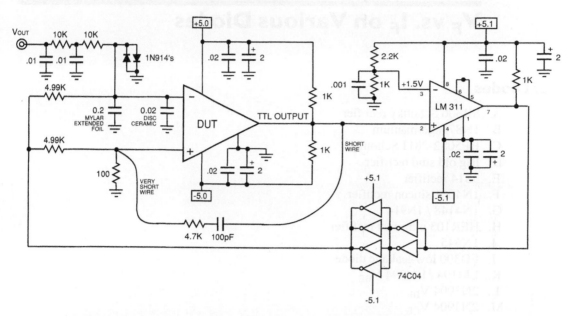

Figure D.2. In practice, you need precision output levels as shown here.

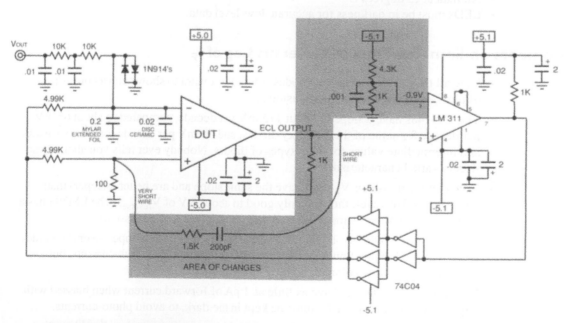

Figure D.3. The circuit in Figure D.2 is for TTL-type comparators; for ECL-output comparators, this circuit is suitable.

Appendix E

V_F vs. I_F on Various Diodes

List of Diodes

A. SR306 Schottky rectifier
B. 1N87G germanium
C. HP5082-2811 Schottky
D. Big old stud rectifier
E. 3S14 rectifier
F. 1N4001 silicon rectifier
G. 1N4148 / 1N914
H. HER103 ultra-fast rectifier
I. 1N645
J. FD300 low-leakage diode
K. LM194 / LM394 V_{BE}
L. 2N3904 V_{BE}
M. 2N3904 V_{CB}
N. LM3046 V_{BE}
O. 4N28 diode, pins 1 and 2.
P. Red LED
Q. Green, yellow, or super-bright red LED.

Notes

- FD200 and FD600 are similar to the 1N4148 curve.
- Transistors have collector shorted to base, for curves K, L, N.
- All data at 25 degrees C.
- LEDs must be in darkness for accurate low-level data.

Comments on Semi-log Plots of V_F versus Log of I_F

- Note all the different slopes of diodes! I wouldn't want to show such a confusing set of plots, except life really is confusing . . .

- The 1N4148s have a slope of about 115 mV per decade, compared to about 65 mV per decade for some of the Schottky diodes, and 60 mV per decade for the transistors, and intermediate values for other types of diodes. Nobody ever tells you about these widely varied characteristics!

- Note that the transistor V_{BE}s all have the same slope and are better (steeper) than most diodes. However, they are only good to about 6 V of $V_{reverse}$. (The LM394 has a built-in base-emitter clamp diode and thus cannot be reverse-biassed.)

- Note that the 2N3904's c-b junction (curve M) has an inferior slope—worse conductance than the b-e junction. Consequently it has a higher V_f at high currents, but worse leakage at low voltages.

- Note that a red LED may have as little as 1 pA of forward current when biassed with 0.6 V forward! But the LEDs must be kept in the dark, to avoid photo-currents.

- Note, I started measuring some of these V_fs with a curve tracer. I realized later that the V_fs were wrong—the curve-tracer was badly out of calibration. I black-flagged it

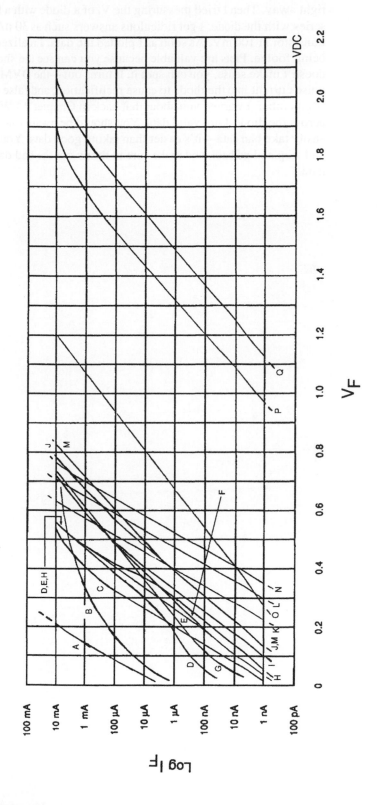

Figure E.1. Semilog plot of V_F versus I_F for various diodes.

right away. Then I tried measuring the V$_f$ of a diode with a battery-powered DVM in series with the diode. I got ridiculous answers such as 30 nA of leakage on a good transistor at 100 mV. As soon as I plotted the data, I realized I was on the verge of being fooled. Plots are valuable because you can see the shape of a curve, and if it doesn't make sense, you can spot it. It turns out—the DVM was pumping enough AC noise current into the diode to cause rectification, and false values of current. (Remember, I warned you about that back in Chapter 2.) When I then put 0.47 μF across the diode, I got valid data. You should be aware that you, as well as I, can easily take bad data—it's easier than taking good data. You just have to be suspicious and stop and go back and make sure to throw out the bad data, and then get valid data.

APPENDIX F
How to Get the Right Information From a Data Sheet

Not All Data Sheets Are Created Alike, and False Assumptions Could Cost an Engineer Time and Money

By Robert A. Pease

When a new product arrives in the marketplace, it hopefully will have a good, clear data sheet with it.

The data sheet can show the prospective user how to apply the device, what performance specifications are guaranteed and various typical applications and characteristics. If the data-sheet writer has done a good job, the user can decide if the product will be valuable to him, exactly how well it will be of use to him and what precautions to take to avoid problems.

SPECIFICATIONS

The most important area of a data sheet specifies the characteristics that are guaranteed—and the test conditions that apply when the tests are done. Ideally, all specifications that the users will need will be spelled out clearly. If the product is similar to existing products, one can expect the data sheet to have a format similar to other devices.

But, if there are significant changes and improvements that nobody has seen before, then the writer must clarify what is meant by each specification. Definitions of new phrases or characteristics may even have to be added as an appendix.

For example, when fast-settling operational amplifiers were first introduced, some manufacturers defined settling time as the time after slewing before the output finally enters and stays within the error-band; but other manufacturers included the slewing time in their definition. Because both groups made their definitions clear, the user was unlikely to be confused or misled.

However, the reader ought to be on the alert. In a few cases, the data-sheet writer is playing a specsmanship game, and is trying to show an inferior (to some users) aspect of a product in a light that makes it look superior (which it may be, to a couple of users).

GUARANTEES

When a data sheet specifies a guaranteed minimum value, what does it mean? An assumption might be made that the manufacturer has actually tested that specification and has great confidence that no part could fail that test and still be shipped. Yet that is not always the case.

For instance, in the early days of op amps (20 years ago), the differential-input impedance might have been guaranteed at 1 MΩ—but the manufacturer obviously did not measure the impedance. When a customer insisted, "I have to know how you measure this impedance," it had to be explained that the impedance was not measured, but that the base current was. The correlation between I_b and Z_{in} permitted the substitution of this simple dc test for a rather messy, noisy, hard-to-interpret test.

Reprinted by permission from Electronic Engineering Times.

Every year, for the last 20 years, manufacturers have been trying to explain, with varying success, why they do not measure the Z_{in} *per se,* even though they do guarantee it.

In other cases, the manufacturer may specify a test that can be made only on the die as it is probed on the wafer, but cannot be tested after the die is packaged because that signal is not accessible any longer. To avoid frustrating and confusing the customer, some manufacturers are establishing two classes of guaranteed specifications:

- The tested limit represents a test that cannot be doubted, one that is actually performed directly on 100 percent of the devices, 100 percent of the time.

- The design limit covers other tests that may be indirect, implicit or simply guaranteed by the inherent design of the device, and is unlikely to cause a failure rate (on that test), even as high as one part per thousand.

Why was this distinction made? Not just because customers wanted to know which specifications were guaranteed by testing, but because the quality-assurance group insisted that it was essential to separate the tested guarantees from the design limits so that the AQL (assurance-quality level) could be improved from 0.1 percent to down below 100 ppm.

Some data sheets guarantee characteristics that are quite expensive and difficult to test (even harder than noise) such as long-term drift (20 ppm or 50 ppm over 1,000 hours).

The data sheet may not tell the reader if it is measured, tested or estimated. One manufacturer may perform a 100-percent test, while another states, "Guaranteed by sample testing." This is not a very comforting assurance that a part is good, especially in a critical case where only a long-term test can prove if the device did meet the manufacturer's specification. If in doubt, question the manufacturer.

TYPICALS

Next to a guaranteed specification, there is likely to be another in a column labeled "typical".

It might mean that the manufacturer once actually saw one part as good as that. It could indicate that half the parts are better than that specification, and half will be worse. But it is equally likely to mean that, five years ago, half the parts were better and half worse. It could easily signify that a few parts might be slightly better, and a few parts a lot worse; after all, if the noise of an amplifier is extremely close to the theoretical limit, one cannot expect to find anything much better than that, but there will always be a few noisy ones.

If the specification of interest happens to be the bias current (I_b) of an op amp, a user can expect broad variations. For example, if the specification is 200 nA maximum, there might be many parts where I_b is 40 nA on one batch (where the beta is high), and a month later, many parts where the I_b is 140 nA when the beta is low.

Absolute Maximum Ratings (Note 11)

If Military/Aerospace specified devices are required, please contact the National Semiconductor Sales Office/Distributors for availability and specifications.

Supply Voltage	+35V to −0.2V
Output Voltage	+6V to −1.0V
Output Current	10 mA
Storage Temperature,	
TO-46 Package	−76°F to +356°F
TO-92 Package	−76°F to +300°F

Lead Temp. (Soldering, 4 seconds)	
TO-46 Package	+300°C
TO-92 Package	+260°C
Specified Operating Temp. Range (Note 2)	
	T_{MIN} to T_{MAX}
LM34, LM34A	−50°F to +300°F
LM34C, LM34CA	−40°F to +230°F
LM34D	+32°F to +212°F

DC Electrical Characteristics (Note 1, Note 6)

Parameter	Conditions	LM34A Typical	LM34A Tested Limit (Note 4)	LM34A Design Limit (Note 5)	LM34CA Typical	LM34CA Tested Limit (Note 4)	LM34CA Design Limit (Note 5)	Units (Max)
Accuracy (Note 7)	$T_A = +77°F$	±0.4	±1.0		±0.4	±1.0		°F
	$T_A = 0°F$	±0.6			±0.6		±2.0	°F
	$T_A = T_{MAX}$	±0.8	±2.0		±0.8	±2.0		°F
	$T_A = T_{MIN}$	±0.8	±2.0		±0.8		±3.0	°F
Nonlinearity (Note 8)	$T_{MIN} \le T_A \le T_{MAX}$	±0.35		±0.7	±0.30		±0.6	°F
Sensor Gain (Average Slope)	$T_{MIN} \le T_A \le T_{MAX}$	+10.0	+9.9, +10.1		+10.0		+9.9, +10.1	mV/°F, min mV/°F, max
Load Regulation (Note 3)	$T_A = +77°F$	±0.4	±1.0		±0.4	±1.0		mV/mA
	$T_{MIN} \le T_A \le T_{MAX}$ $0 \le I_L \le 1$ mA	±0.5		±3.0	±0.5		±3.0	mV/mA
Line Regulation (Note 3)	$T_A = +77°F$	±0.01	±0.05		±0.01	±0.05		mV/V
	$5V \le V_S \le 30V$	±0.02		±0.1	±0.02		±0.1	mV/V
Quiescent Current (Note 9)	$V_S = +5V, +77°F$	75	90		75	90		µA
	$V_S = +5V$	131		160	116		139	µA
	$V_S = +30V, +77°F$	76	92		76	92		µA
	$V_S = +30V$	132		163	117		142	µA
Change of Quiescent Current (Note 3)	$4V \le V_S \le 30V, +77°F$	+0.5	2.0		0.5	2.0		µA
	$5V \le V_S \le 30V$	+1.0		3.0	1.0		3.0	µA
Temperature Coefficient of Quiescent Current		+0.30		+0.5	+0.30		+0.5	µA/°F
Minimum Temperature for Rated Accuracy	In circuit of *Figure 1*, $I_L = 0$	+3.0		+5.0	+3.0		+5.0	°F
Long-Term Stability	$T_j = T_{MAX}$ for 1000 hours	±0.16			±0.16			°F

Note 1: Unless otherwise noted, these specifications apply: $-50°F \le T_j \le +300°F$ for the LM34 and LM34A; $-40°F \le T_j \le +230°F$ for the LM34C and LM34CA; and $+32°F \le T_j \le +212°F$ for the LM34D. $V_S = +5$ Vdc and $I_{LOAD} = 50$ µA in the circuit of *Figure 2;* +6 Vdc for LM34 and LM34A for 230°F $\le T_j \le$ 300°F. These specifications also apply from +5°F to T_{MAX} in the circuit of *Figure 1*.

Note 2: Thermal resistance of the TO-46 package is 792°F/W junction to ambient and 43°F/W junction to case. Thermal resistance of the TO-92 package is 324°F/W junction to ambient.

Note 3: Regulation is measured at constant junction temperature using pulse testing with a low duty cycle. Changes in output due to heating effects can be computed by multiplying the internal dissipation by the thermal resistance.

Note 4: Tested limits are guaranteed and 100% tested in production.

Note 5: Design limits are guaranteed (but not 100% production tested) over the indicated temperature and supply voltage ranges. These limits are not used to calculate outgoing quality levels.

Note 6: Specification in **BOLDFACE TYPE** apply over the full rated temperature range.

Note 7: Accuracy is defined as the error between the output voltage and 10 mV/°F times the device's case temperature at specified conditions of voltage, current, and temperature (expressed in °F).

Note 8: Nonlinearity is defined as the deviation of the output-voltage-versus-temperature curve from the best-fit straight line over the device's rated temperature range.

Note 9: Quiescent current is defined in the circuit of *Figure 1*.

Note 10: Contact factory for availability of LM34CAZ.

Note 11: Absolute Maximum Ratings indicate limits beyond which damage to the device may occur. DC and AC electrical specifications do not apply when operating the device beyond its rated operating conditions (see Note 1).

A Point-By-Point Look

Let's look a little more closely at the data sheet of the National Semiconductor LM34, which happens to be a temperature sensor.

Note 1 lists the nominal test conditions and test circuits in which all the characteristics are defined. Some additional test conditions are listed in the column "Conditions", but Note 1 helps minimize the clutter.

Note 2 gives the thermal impedance, (which may also be shown in a chart or table).

Note 3 warns that an output impedance test, if done with a long pulse, could cause significant self-heating and thus, error.

Note 6 is intended to show which specs apply at all rated temperatures.

Note 7 is the definition of the "Accuracy" spec, and Note 8 the definition for non-linearity. Note 9 states in what test circuit the quiescent current is defined. Note 10 indicates that one model of the family may not be available at the time of printing (but happens to be available now), and Note 11 is the definition of Absolute Max Ratings.

* Note—the "4 seconds" soldering time is a new standard for plastic packages.

** Note—the wording of Note 11 has been revised—this is the best wording we can devise, and we will use it on all future datasheets.

APPLICATIONS

Another important part of the data sheet is the applications section. It indicates the novel and conventional ways to use a device. Sometimes these applications are just little ideas to tweak a reader's mind. After looking at a couple of applications, one can invent other ideas that are useful. Some applications may be of no real interest or use.

In other cases, an application circuit may be the complete definition of the system's performance; it can be the test circuit in which the specification limits are defined, tested and guaranteed. But, in all other instances, the performance of a typical application circuit is not guaranteed, it is only typical. In many circumstances, the performance may depend on external components and their precision and matching. Some manufacturers have added a phrase to their data sheets:

"Applications for any circuits contained in this document are for illustration purposes only and the manufacturer makes no representation or warranty that such applications will be suitable for the use indicated without further testing or modification."

In the future, manufacturers may find it necessary to add disclaimers of this kind to avoid disappointing users with circuits that work well, much of the time, but cannot be easily guaranteed.

The applications section is also a good place to look for advice on quirks—potential drawbacks or little details that may not be so little when a user wants to know if a device will actually deliver the expected performance.

For example, if a buffer can drive heavy loads and can handle fast signals cleanly (at no load), the maker isn't doing anybody any favors if there is no mention that the distortion goes sky-high if the rated load is applied.

Another example is the application hint for the LF156 family:

"Exceeding the negative common-mode limit on either input will cause a reversal of the phase to output and force the amplifier output to the corresponding high or low state. Exceeding the negative common-mode limit on both inputs will force the amplifier output to a high state. In neither case does a latch occur, since raising the input back within the common-mode range again puts the input stage and, thus the amplifier, in a normal operating mode."

That's the kind of information a manufacturer should really give to a data-sheet reader because no one could ever guess it.

Sometimes, a writer slips a quirk into a characteristic curve, but it's wiser to draw attention to it with a line of text. This is because it's better to make the user sad before one gets started, rather than when one goes into production. Conversely, if a user is going to spend more than 10 minutes using a new product, one ought to spend a full five minutes reading the entire data sheet.

FINE PRINT

What other fine print can be found on a data sheet? Sometimes the front page may be marked "advance" or "preliminary." Then on the back page, the fine print may say something such as:

"This data sheet contains preliminary limits and design specifications. Supplemental information will be published at a later date. The manufacturer reserves the right to make changes in the products contained in this document in order to improve design or performance and to supply the best possible products. We also assume no responsibility for the use of any circuits described herein, convey no license under any patent or other right and make no representation that the circuits are free from patent infringement."

In fact, after a device is released to the marketplace in a preliminary status, the engineers love to make small improvements and upgrades in specifications and characteristics, and hate to degrade a specification from its first published value—but occasionally that is necessary.

Another item in the fine print is the manufacturer's telephone number. Usually it is best to refer questions to the local sales representative or field-applications engineer, because they may know the answer or they may be best able to put a questioner in touch with the right person at the factory.

Occasionally, the factory's applications engineers have all the information. Other times, they have to bring in product engineers, test engineers or marketing people. And sometimes the answer can't be generated quickly—data have to be gathered, opinions solidified or policies formulated before the manufacturer can answer the question. Still, the telephone number is the key to getting the factory to help.

ORIGINS OF DATA SHEETS

Of course, historically, most data sheets for a class of products have been closely modeled on the data sheet of the forerunner of that class. The first data sheet was copied to make new versions.

That's the way it happened with the UA709 (the first monolithic op amp) and all its copies, as well as many other similar families of circuits.

Even today, an attempt is made to build on the good things learned from the past and add a few improvements when necessary. But, it's important to have real improvements, not just change for the sake of change.

So, while it's not easy to get the format and everything in it exactly right to please everybody, new data sheets are continually surfacing with new features, applications ideas, specifications and aids for the user. And, if the users complain loudly enough about misleading or inadequate data sheets, they can help lead the way to change data sheets. That's how many of today's improvements came about—through customer demand.

Who writes data sheets? In some cases, a marketing person does the actual writing and engineers do the checking. In other companies, the engineer writes, while marketing people and other engineers check. Sometimes, a committee seems to be doing the writing. None of these ways is necessarily wrong.

For example, one approach might be: The original designer of the product writes the data sheet (inside his head) at the same time the product is designed. The concept here is, if one can't find the proper ingredients for a data sheet—good applications, convenient features for the user and nicely tested specifications as the part is being designed—then maybe it's not a very good product until all those ingredients are completed. Thus, the collection of raw materials for a good data sheet is an integral part of the design of a product. The actual assembly of these materials is an art which can take place later.

WHEN TO WRITE DATA SHEETS

A new product becomes available. The applications engineers start evaluating their application circuits and the test engineers examine their production test equipment.

But how can the users evaluate the new device? They have to have a data sheet—which is still in the process of being written. Every week, as the data sheet writer tries to polish and refine the incipient data sheet, other engineers are reporting, "These spec limits and conditions have to be revised," and, "Those application circuits don't work like we thought they would; we'll have one running in a couple of days." The marketing people insist that the data sheet must be finalized and frozen right away so that they can start printing copies to go out with evaluation samples.

These trying conditions may explain why data sheets always seem to have been thrown together under panic conditions and why they have so many rough spots. Users should be aware of the conflicting requirements: Getting a data sheet "as completely as possible" and "as accurately as possible" is compromised if one wants to get the data sheet "as quickly as possible."

The reader should always question the manufacturer. What are the alternatives? By not asking the right question, a misunderstanding could arise; getting angry with the manufacturer is not to anyone's advantage.

Robert Pease has been staff scientist at National Semiconductor Corp., Santa Clara, Calif., for eleven years. He has designed numerous op amps, data converters, voltage regulators and analog-circuit functions.

Appendix G

More on SPICE

Recently I was down in New Orleans at one of the IEEE conferences—the International Symposium on Circuits and Systems. The keynote speaker, Professor Ron Rohrer from Carnegie-Mellon University, commented thoughtfully about many aspects of education for engineers. But what he said that really stunned me was his observation that "in the era of SPICE, nobody designs on the back of envelopes any more." Ouch. It is becoming more and more true that young (or, lazy?) engineers cannot do much designing without some computers or high-powered calculators. SPICE just happens to be one of my pet peeves, and I will start gnawing on its ankles today.

Now, I have always been a friend of analogies, analogues, analogs, similes, models and metaphors. When I worked at George A. Philbrick Researches, the company's motto was, "The analog way is the model way." In those days we sold some analog computers, even though that part of the business was shrinking and the popularity of the op amp was on the rise. But we all tried to follow the party line, that analog computation was a serious business—and it still is, although as a percentage of the electronics business, it has shrunk to a tiny fraction. Still, there are many times where a little analog computation is exactly the right thing, and someday I will expound on that . . .

But, SPICE (Simulation Program with Integrated Circuit Emphasis) is a rather popular and powerful tool these days, and almost everybody finds it useful to some extent. I remember when my old boss, Tim Isbell, showed me how to use it—and then we spent half a day horsing around because it said we had a 72-V forward voltage across a diode, but there was no current through the diode. I will emphasize today just a few of the basic problems with SPICE.

The first main problem is that people tend to trust its answers, as they trust most computers, long after the reason to trust it should have evaporated. I have come very close to fistfights and screaming contests, when a person claims that such-and-such an answer is obviously right because SPICE gave it to him. Conversely, I normally try to avoid working with SPICE unless I can run a calibration program on it, a sanity check, so it gives me an answer that makes sense.

This is much like the old days of the slide rule: You couldn't use the slide rule unless you already knew approximately what the answer was. It's not like a calculator where the decimal place is provided on a platter—you have to provide your own decimal place. In other words, you are forced to be a pretty good engineer before you even pick up your slide rule, or your analog computer.

But people who use SPICE are often buffaloed or fooled by any absurd kind of answer. So, trusting your computer seems to be one of the new trends, which I want to see quashed. It's too easy to find (weeks later) that the computer told you a lie, because the data typed in had a typo error, or a monumental goof. Now, never let it be said that RAP recommends you use analog computers or breadboards instead of SPICE because analog computers do not make errors. SPICE lies, but analog computers do not? Oh, please, don't say that: Analog computers lie, too, and so do bread-

Originally published in *Electronic Design*, November and December 1990.

boards, but I like them because they often offer a greater insight and understanding as to what's really going on, so if you survive their problems, you are smart enough to keep out of other kinds of trouble. But, that's just a bunch of philosophical stuff.

The thing that makes me nervous about SPICE is that it was largely designed by a group of grad students (Laurence Nagel and others at Berkeley) back in '73. Now, when you find a problem, a discrepancy, a glitch, a flaw, an error that seems to be built into SPICE, can you go back to the people who designed it? Hardly. There is no continuity. There *are* some people who claim to "support" SPICE, but I don't always agree with their statements.

My biggest gripe with SPICE is its lack of convergence. The ordinary SPICE 2G6 has all sorts of problems, even if you don't use FETs. (We find that FETs usually make the convergence situation *really* unhappy.) For example, one time I had a moderate-sized circuit with about 33 bipolar transistors, and it didn't converge well. Then all of a sudden, one day it started to converge beautifully and quickly. I was so impressed, I backed up to find the "scene of the crime." I tried to duplicate all the changes I had made since I last had problems. It finally turned out that I had an unused resistor and an un-used capacitor each tied from one point to ground. Nothing else was connected to that point. Originally they were "commented out" by an asterisk. But, at one point, I deleted the asterisk, and the useless R and C were dropped into the circuit—and they just happened to make the convergence a lot better. When I removed the R and C, things got worse again.

This lead me to appreciate two things: That the convergence is a lot more fragile than we suspect, and that we may be able to randomly throw useless resistors into a circuit, and sometimes they could improve the convergence. In other words, if you have a circuit that shows bad convergence, the computer might have a subroutine to randomly sprinkle a few resistors into the circuit and see if that helps—a kind of "autoconverge" scheme. At present we are still working on this, but it may be a useful approach. This concept is not *too* surprising if you have ever heard that the convergence of a circuit may be improved or degraded *depending on the names and numbers you call the nodes of the circuit.* If you swap a couple nodes' numbers, and things get better (or worse)—doesn't that make you nervous?? Or, at least, *suspicious?*

Another serious problem I had with SPICE was when I ran some simple transient tests—triangle waves—on the collector of a transistor. I ramped the V_c up and down from +5 V to +15 V, back and forth, and ran several tests. Then I added some complicating factors, so I wanted to look at the circuit for the first 202 μs of a 10-kHz triangle wave. Namely, after the first two cycles of the triangle wave, I decided to look at the collector current ($I = C \times dV/dt$) of the transistor at t = 201 μs. I got my answer printed and plotted, and it did not make any sense. I studied the whole circuit, and I used every troubleshooting technique I could think of, and it did not make any sense. The current through the 1 pF of C_{bc} was not 0.2 μA, but 5 μA. How could that be?

After several hours, I finally decided to look at the incoming waveform. I had commanded it to go back and forth from 5 V to 15 V, at the rate of 50 μs per each ramp, so I *knew* what it had to be doing. But when I looked at t = 201 μs, the dV/dt had suddenly increased from 0.2 V/μs to 5 V/μs. It turned out that because I had commanded the PLOT mode to stop at 202 μs, the transient generator had decided to go from 15 V to 5 V, not in the time from 200 to 250 μs, but in the span from 200 to 202 μs. The dV/dt speeded up by a factor of 25, without being asked to, for a completely unexpected reason. Nothing in anything I had ever seen about SPICE, nothing my friends had ever heard, would lead you to expect this. In fact, SPICE sort of encourages you to look at the waveforms any time you want—it offers a sort of "infinitely versatile expanded-scale oscilloscope" and if it has a dV/dt that suddenly changes, well that is quite a surprise.

So, I immediately wrote an open message to all my friends at NSC, warning them about this potential problem; and now I am writing about this to warn all my friends everywhere. So, these are just some of the reasons I am not enthusiastic about SPICE. It's goofed me up, me and my friends, too many times.

My boss points out that it's not necessarily true that all kinds of SPICE have such bad problems with convergence or bad computations or spurious signals. And that may be so. If somebody who knows all about W-SPICE or J-SPICE wants to write in and assure me that *his* SPICE will never do that, well, that is fine by me. But, meanwhile, don't get me wrong—I don't *hate* these digital computers. *They* hate me; I *despise* them.

The other day I was standing out in the rain, talking with a design engineer from the East Coast. He said all the other engineers at his company ridicule him because they rely on SPICE, while he depends on the breadboards he builds. There's just one hitch: his circuits work the first time, and theirs don't. To add insult to injury, his boss forces him to help his colleagues get their circuits working, since he has so much time left over. I said that sounds pretty good to me, so long as his boss remembers who is able to get out the circuits, when it comes to doing reviews for all the guys.

This guy gave me a tip: Don't design a circuit in SPICE with 50 Ω resistors. Use 50.1 Ω. It converges better. H'mmmm. That sounds intriguing.

Right now I am struggling with a SPICE model of a circuit. Not of a new circuit, but of an old circuit: The band-gap reference of the old LM331 which I put into production back in '77. It's a good thing I put it into production before we got SPICE, because if I had first run this through SPICE, I would have been pretty discouraged. SPICE says this circuit has a rotten temperature coefficient and oscillates like a politician. I went back and double-checked the actual silicon circuits. They soar like an angel, have very low TC, and are dead-beat when you bang on them. They have no tendency to oscillate; they do not even ring. So why does SPICE persist in lying to me? Doesn't it realize I will break its back, for the impertinence of lying to the Czar of Bandgaps?? The SPICE and CAD experts around here tell me, "Oh, you must have bad models." I've been told that before, when I was right and the experts were absolutely wrong. (I mean, how can a single FET oscillate at 400 kHz?? With the help of two resistors...) More on this topic, later.

I've already gotten several letters from readers who have asked, "How about all these new models for op amps? Won't they lead linear designers in a new direction?" My replies to them start out by covering a couple examples of old macro-models of op amps that have been raising questions for over a dozen years.

A guy calls up and asks me, "What is the maximum DC voltage gain on an LM108?" I reply, "Well, it's 40,000 min, but a lot of them run 300,000 or 500,000, and some of them are as high as 3 or 4 million." The customer sighs, "Oh, that's terrible...." When I ask why it's terrible, he explains that when the gain gets high, the gain bandwidth will get so high that it will be impossible to make a stable loop, if the gain bandwidth gets up to dozens or hundreds of megahertz.

Sigh. I sit down and explain that there is no correlation between the DC gain and its spread, compared to the GBW Product and its spread. The guy says, "Oh, I read in a book somewhere that there's good correlation, because the first pole is constant." I tell him to throw out the book, or at least X out those pages, because the first pole is *not* at a constant frequency.

These days, I read that several op-amp companies are giving away free SPICE models. What do I think of these models? Well, on a *typical* basis, I have read that some of these models are pretty good, as in several *typical* situations, they slew and settle (and ring just a little, as real op amps do) and have as good accuracy as a real *typical* op amp and its feedback resistors. Maybe in a few years the slow ones will be trustworthy. But I don't think you can get very good results from the fast ones. Why? PC layout strays. Enough said.

And besides, how good are those models if you ask their makers? Are the models guaranteed to give such a good representation of reality that if SPICE gives good results, the op amps are guaranteed to work? Well..........no, not exactly. In fact, from what I have read, none of the op-amp models are guaranteed for anything. The only thing they can do, "guaranteeably," is to give a customer *something* when he begs for SPICE models. It's guaranteed to make the customer go away happy and to keep him busy for a while. But it's not guaranteed to make him happy in the long run. Because the performance of high-speed op amps and precision circuits depends so critically on the layout, and on the resistors and capacitors, that the model itself is almost irrelevant.

Now some people might say, "How does Pease dare to say that?"

It's easy. I haven't got any SPICE models of my op amps to give away. Not at this time. And if I did, or when I do, I won't be able to guarantee them either. At best, I may be able to say, "If you are a good engineer and use these models as a tool to pioneer some experiments that are inconvenient to test on the breadboard, you may find these models are helpful, so long as you then check it out on your breadboard to confirm the circuit. For example, you can use SPICE to 'measure' some voltages or currents that are so small and delicate that you really could not measure them with a scope or buffered probe or current probe, not in the real world. *But* if you try to rely solely on these models, without breadboarding, they will not tell you the whole story, and your crutches will collapse, sooner or later, and you can't say I didn't warn you."

I showed this to Bettina Briz in Amplifier Marketing, and she said, "Bob, you can't say *that*." I said, "Oh, tell me where I have said anything that is untrue, and I will fix it." She admitted that what I had said probably was .... quite true. Then I said, "Well, why try to soft-pedal the truth, and pretend that you can trust computers all the time? Wouldn't that be a disservice to our customers?" And Bettina replied, "When *we* have models, we'll educate our users—*we'll* point out when you can trust the models, and when you shouldn't. So—after that—are we in disagreement?" Well—maybe we did agree after all.

At present, we have a small library of op-amp models released with Analogy (Beaverton, Oregon 97075).[1] They are only level I models (low precision), and while we have made some progress on good-precision ones (level II), they are not released yet. These are "behavioral models" rather than SPICE models, and we think they have several advantages over SPICE models. There are some min/typ/max specifications that pretty much correspond to data sheet limits. If you use them wisely, they may be helpful—subject to the conditions I listed in the previous paragraph. These models are not free. They are not even cheap. But we think they are worth what you pay for them. Still, they are not guaranteed.

Now, seriously, where can you get a model of a transistor that is guaranteed? To run under all conditions? I don't think you can beg or steal or borrow or buy a model of a transistor that is guaranteed.[2] Or of a capacitor. Or even of a resistor.

But I can guarantee that every op amp you can buy or make has some characteris-

1. Analogy Inc., P.O. Box 1669, Beaverton, Oregon 97075, (503) 626-9700.

2. For information on guaranteed models for CMOS transistors, inquire with James Smith, Semiconductor Physics, Inc., 639 Meadow Grove Place, Escondido, California 92027-4236 (619) 741-3360.

tics that cannot be absolutely modeled by any computer model. If you happen to depend on that feature, *or* the absence of that feature, it is only a matter of time before you get in trouble.

I will also guarantee that just because you made one breadboard, and it works well, you cannot put that circuit into production and get 1000 units in a row to work well, *unless* you are a smart engineer and design the circuit "properly" and do your worst-case design studies, and plan for well-behaved frequency response, etc. And I think that is true, no matter where you buy your op amps. What's new? What color is the king's new underwear? Dirty grey, same as everybody else's.

I was at an evening session at the IEEE Bipolar Circuits and Technology Meeting in Minneapolis recently. Several companies that sell CAD tools had done some serious work to analyze the circuit for a 12-bit A/D converter. Even the ones that had only a little time to put in were able to show that macro-models were feasible and effective as a way to do good analysis while saving computing time—that was the primary objective of the study. But even the ones that put in the *most* time at analysis did not recognize (or did not comment about it) that the noise of the reference and the comparator were rather large, and you could not achieve 12-bit resolution without slowing down the response a lot more than you would have to do otherwise (for a circuit where you did not have to consider noise).

If a good designer of A/D converters had these tools, and he knew where to look for noise, or where to insert lead inductance, or extra substrate capacitances, he might use some of these CAD tools to help him design a better ADC. But if he just believed what the computer told him, he would probably be badly fooled.

Once a customer called me up and asked me how to get my LM108s to stop oscillating in his circuit. He explained it was a simulated LM108 with some simulated feedback resistors, and simulated switches and filters. Oh. H'mmmm. I asked if he had made up a breadboard, and, did it oscillate? He said he had made it and it did not oscillate. H'mmmmm. I asked him, "If you built up a breadboard and a computer model, and the real breadboard oscillated, but the computer did not, you wouldn't be calling up to complain about the computer, would you?" He stopped and thought about it. He cogitated for a while. He said "I'll call you back." And he hung up. And he never did call back. I mean, what would *you* do?

Appendix H

Pease's Troubleshooting Articles as Originally Published in *EDN*

Part 1—"Troubleshooting is more effective with the right philosophy," *EDN*, January 5, 1989, p. 147.

Part 2—"The right equipment is essential for effective troubleshooting," *EDN*, January 19, 1989, p. 166.

Part 3—"Troubleshooting gets down to the component level," *EDN*, February 2, 1989, p. 175.

Part 4—"A knowledge of capacitor subtleties helps solve capacitor-based troubles," *EDN*, February 16, 1989, p. 127.

Part 5—"Follow simple rules to prevent material and assembly problems," *EDN*, March 2, 1989, p. 159.

Part 6—"Active-component problems yield to painstaking probing," *EDN*, August 3, 1989, p. 127.

Part 7—"Rely on semiconductor basics to identify transistor problems," *EDN*, August 17, 1989, p. 129.

Part 8—"Keep a broad outlook when troubleshooting op-amp circuits," *EDN*, September 1, 1989, p. 131.

Part 9—"Troubleshooting techniques quash spurious oscillations," *EDN*, September 14, 1989, p. 151.

Part 10—"The analog/digital boundary needn't be a never-never land," *EDN*, September 28, 1989, p. 145.

Part 11—"Preside over power components with design expertise," *EDN*, October 12, 1989, p. 177.

Part 12—"Troubleshooting series comes to a close," *EDN*, October 26, 1989, p. 171.

Part 13—"Pease's pointers rouse readers: Letters to Bob," *EDN*, May 10, 1990, p. 119.

Index

References to figures are printed in boldface type.

Notes for the U.K. Reader

The following notes are provided by the publisher specifically for readers in the U.K.:

1. The prices quoted in dollars ($) should of course be converted to sterling at the current rate of exchange (presently about $2 to the pound). It should, however, be noted that prices in the U.K. are usually somewhat in excess of the straightforward converted sum.

2. The symbol "#" is used to denote "number." Hence, "resistor #20" is equivalent to "R20," and so on.

3. Various items of test equipment of U.S. origin (e.g., Hewlett Packard, Fluke, etc.) are available in the U.K. However, manufacturers such as Philips, Marconi, Gould-Advance, and Farnell can, in many cases, provide equipment which will operate to a comparable specification.

4. The variable autotransformers (variacs) mentioned in Chapter 2 can be obtained from several suppliers in the U.K. including RS Components (types rated at 0.5 A, 2 A, and 8 A having Stock Codes 207–936, 207–611, and 207–914 respectively) and Farnell Electronic Components (a larger range with various types rated at between 0.5 A and 12.5 A and manufactured by Zenith and Claude Lyons).

5. The excellent Heathkit HD-1250 dip meter (also mentioned in Chapter 2) is available in kit form in the U.K. from Maplin Professional Supplies (Stock Code: HK23A). Maplin can also provide a cheaper, ready-built dip meter (Stock Code: YN48C) which covers an almost identical frequency range.

Addresses of U.K. suppliers:

Farnell Electronic Components
Canal Road
Leeds
West Yorkshire
LS12 2TU
Telephone: 0532-636311

RS Components
P.O. Box 99
Corby
Northants
NN17 9RS
Telephone: 0536-201234

Maplin Professional Supplies
P.O. Box 777
Rayleigh
Essex
SS6 8LU
Telephone: 0702-552961

Also from the EDN Series for Design Engineers,
published by Butterworth-Heinemann

Analog Circuit Design
Art, Science, and Personalities

edited by Jim Williams

Coping with analog circuit design is a lot easier if you have an experienced "analog wizard" to guide you through the design process. Jim Williams is such a wizard, and in this book he's assembled 23 other masters of the analog art to share their experience, knowledge, insights, and often wit.

Analog Circuit Design: Art, Science, and Personalities is far more than just another tutorial or reference guide—it's a tour through the world of analog design, combining theory and applications with the philosophies behind the design process. You'll read how leading analog circuit designers approach problems, and you'll gain insight into their thought processes as they develop solutions to those problems. If you work with analog circuits, or want to gain new understanding of what can be a complex subject, this is one book you must have.

Look for this book at your technical bookstore, or order directly from Butterworth-Heinemann.

1991, 389 pages, ISBN 0-7506-9166-2